Molecular Biology Biochemistry and Biophysics

33

Erich Heinz

Electrical Potentials in Biological Membrane Transport

With 15 Figures

Springer-Verlag
Berlin Heidelberg New York 1981

Professor Dr. ERICH HEINZ
Department of Physiology
Cornell University
Medical College
1300 York Avenue
New York, NY 10021/USA

ISBN 3-540-10928-5 Springer-Verlag Berlin Heidelberg New York
ISBN 0-387-10928-5 Springer-Verlag New York Heidelberg Berlin

Library of Congress Cataloging in Publication Data. Heinz, Erich, 1912-, Electric potentials in biological membrane transport. (Molecular biology, biochemistry, and biophysics ; v. 33). Bibliography: p. Includes index. 1. Biological transport, Active. 2. Membranes (Biology)--Electric properties. I. Title. II. Series.
QH509.H43 574.87'5 81-9301. AACR2.

Printed in Germany.

Typesetting and Offsetprinting: Beltz Offsetdruck, Hemsbach/Bergst.
Bookbinding: Brühlsche Universitätsdruckerei, Giessen.
2131/3130-543210

Preface

The material of this volume was originally planned to be incorporated in the preceding monograph Mechanics and Energetics of Biological Transport. A separate and coherent treatment of the variety of bioelectrical phenomena was considered preferable, mainly for didactic reasons. Usually, the biologist has to gather the principles of bioelectricity he needs from different sources and on different levels. The present book intends to provide these principles in a more uniform context and in a form adjusted to the problems of a biologist, rather than of a physicist or electrical engineer. The main emphasis is put on the molecular aspect by relating the bioelectrical phenomena, such as the membrane diffusion potentials, pump potentials, or redox potentials, to the properties of the membrane concerned, and, as far as possible, to specific steps of transport and metabolism of ions and nonelectrolytes. Little space is devoted to the familiar and widely used representation of bioelectrical phenomena in terms of electrical networks, of equivalent circuits with batteries, resistances, capacities etc. In order to elucidate the basic principles, the formal treatment is kept as simple as possible, using highly simplified models, based on biological systems. The corresponding equations are derived in two ways: kinetically, i.e. in terms of the Law of Mass Action, as well as energetically, i.e., in terms of Nonequilibrium Thermodynamics. Phase boundary potentials, which usually affect transmembrane potentials more indirectly, are treated rather cursorily, with the mere purpose to give the reader a rough idea of their nature. Also the discussion of the technical problems connected with the measurement of electrical phenomena is reduced to the bare principles, to the extent that they are useful as a basis to study the abundant literature on the various methods offered to assay bioelectrical potentials and currents. It is hoped that this volume will be useful to those biologists who are primarily interested in the biological processes as such, and who often consider electrical phenomena an unwelcome complication.

Summer 1981 ERICH HEINZ

Contents

List of Symbols

A_r	affinity of an overall coupled process
A_{ch}	affinity of a chemical reaction
a_i	activity of an undefined solute i
$\widetilde{a}_i$	electrochemical activity of undefined solute i (= $a_i\xi_i$)
c_i	concentration of undefined solute i
E_{red}	Redox PD
F	Faraday constant
g_i	specific electrical conductance of solute i
J_i	net flow of an indefined solute i per reference unit (e.g., g protein, or unit area) positive in the direction from left (') to right ("), or into a cell or organelle
J_r	rate of overall process, per reference unit
K_m	half-saturation (Michaelis) constant
L_r	phenomenological coefficient for overall process
L_i^u	phenomenological leakage coefficient of solute i
ln	natural logarithm
P	product of biochemical reaction
p	concentration of P
P_i	permeability coefficient of solute i
R	gas constant
R_i	phenomenological resistance coefficient
S	substrate of biochemical reaction
s	concentration of S
T	absolute temperature
u_+, u_-	mobilities of cation and anion, respectively
X_i	osmotic driving force (= $-\Delta\widetilde{\mu}_i$) of solute i
z_i	number of electric charges of ion i

Greek Symbols

Γ	chemical reactivity coefficient [= $\exp\left(\frac{A_{ch}}{RT}\right)$]
δ	thickness of membrane
μ_i	chemical potential of solute i
$\widetilde{\mu}_i$	electrochemical potential of solute i

ν_i	stoichiometric coefficient of solute i (numerical only)
Ψ	electric (membrane) potential
ξ_i	electrochemical activity coefficient $[= \exp(-z_i \frac{F\Delta\Psi}{RT})]$

Abbreviations

EMF	electron motive force
LMA	Law of Mass Action
PD	potential difference
PMF	protonmotive force
TIP	Thermodynamics of Irreversible Processes

Introduction

Almost every living biological membrane maintains an electrical potential difference between the two adjacent solutions. This applies to the cytoplasmic membrane of single cells as well as to tissue membranes such as epithelial and endothelial layers and also to the membranes of subcellular organelles, in particular of mitochondria. Even vesicles formed artificially from biological membranes may under certain conditions show a transient electrical potential.

From the biological potential difference (PD) a weak but steady current can be drawn by suitable electrodes. This current as well as the major part of the PD depend on the metabolic activity of the cells or organelles concerned. Accordingly, complete metabolic inhibition or death eventually causes decay of the potential down to a small residual value, from which current can no longer be drawn.

The biological membrane PD is attributed in part to membrane diffusion potentials across the membranes concerned, and in part to a direct contribution of "electrogenic" ion pumps, as will be discussed in more detail in subsequent chapters. The maintenance of a stationary PD over an appreciable length of time, whatever its origin, requires the continuous operation of ion pumps, which would explain also the metabolic dependence of the PD.

Under normal conditions the biological PD is not always stationary but subject to more or less sudden and temporary changes, which are presumably caused by changes in permeability for certain ions or by variations of the pumping rate. On the other hand, changes in PD from whichever origin may in some systems cause certain alterations of membrane permeability. In general, changes in PD may be either associated with physiological functions of the cell or with pathological perturbations of cellular metabolism.

Electrical potantial differences are part of the electrochemical potential difference of ions across the membrane and hence add to the forces driving the ions across the membranes. They have therefore to be taken into account whenever the movement of an ion is to be related to its conjugate driving force, for instance in order to determine whether the transport of an ion across the membrane is active or passive.

The precise determination of an electrical potantial difference between different solutions poses special difficulties which are discussed in the last section of this book.

The following chapters will develop in more detail the theoretical principles on which the evaluation of electrical observations may be based. As the main emphasis is placed on a basic understanding of these principles rather than on their historic development, the references are to more general treatments only which may help the reader to find more specific literature.

1 Origin of Electrical Potentials

The various kinds of electrical potential that are connected with biological membranes can be divided in various ways. For instance, according to their location we may distinguish between "transmembrane" potentials and "membrane surface" potentials. The former are due to an electric field which penetrates the whole membrane phase and can, therefore, be detected between the two adjacent bulk solutions by the usual electrodes. The latter cannot easily be so detected as they are located at the boundary between the membrane phase and the adjacent bulk solution. One can also divide the potentials from an energetic point of view, namely into "equilibrium" potentials and "nonequilibrium" potentials. The former belong the systems which have reached complete equilibrium or are restained from doing so and cannot therefore serve as a source of free energy, whereas the latter tend to change toward equilibrium, thereby making (under suitable conditions) free energy available.

1.1 Equilibrium Potentials

1.1.1 The Gibbs-Donnan Potential

This obtains if a nonpermeant ion, usually a poly-ion such as protein, is unequally distributed between the two solutions separated by a permselective membrane, which permits certain electrolyte ions to move freely between the two solutions (Donnan, 1911). This movement is restrained by the principle of electroneutrality, which requires that the sum of all cationic charges be equivalent to that of all anionic charges in each solutions, and by the second law of thermodynamics which requires that each permeant ion species moves only down its electrochemical potential gradient.

The relation between ion concentration and electrical PD in the Donnan equilibrium may be illustrated by a simple model: two compartments are separated by a semipermeable membrane. Net water movement through the membranes is assumed to be prevented by the rigidity of the enclosing walls and membrane. The Na^+ salt of polyanion, e.g., protein (X^{n-}), is present on the right side (") only, whereas some $Na^+ Cl^-$ is initially present on each side at equal concentrations (Fig. 1).

It is assumed that the poly-ion, owing to its size and charge, cannot penetrate the membrane, whereas the Na and Cl-ions do so readily. The system may proceed toward equilibrium by the movements of Na^+ and Cl^-, which to maintain electroneutrality must be electrically equivalent to each other.

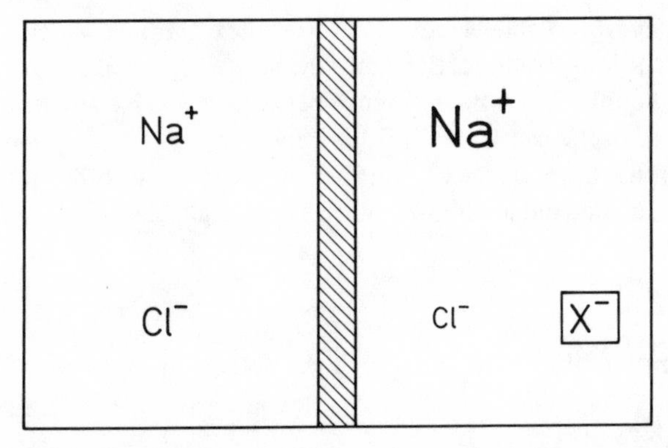

Fig. 1. Donnan distribution of Na Cl between two compartment separated by a membrane (*shaded area*). X^- represents a nonpermeant polyvalent anion. The differences in letter size for Na^+ and Cl^- relate to the differences in the equilibrium concentrations of the ions concerned. For details see text

In equilibrium the difference in electrochemical potential of each ion between the two sides of the membrane has vanished. The electrochemical potential of an ion i, $\tilde{\mu}_i$ is usually sufficiently characterized by the term

$$\tilde{\mu}_i = \tilde{\mu}_i^o (T) + RT \ln \gamma_i c_i + z_i F \Psi$$

$\mu_i^o (T)$ being the standard potential of i at the temperature T, γ_i the activity coefficient, c_i the concentration, z_i the electric valency of i, F the Faraday constant and Ψ the electrical potential. As z is +1 for Na^+ and −1 for Cl^-, the differences in electrochemical potential between the two compartments for these ions, assuming that temperature and activity coefficients are the same, are

$$\Delta\tilde{\mu}_{Na} = RT \ln \frac{[Na]''}{[Na]'} + F\Delta\Psi \tag{1a}$$

$$\Delta\tilde{\mu}_{Cl} = RT \ln \frac{[Cl]''}{[Cl]'} - F\Delta\Psi \tag{1b}$$

As in equilibrium each of these is zero, there will be an electrical potential difference (Donnan potential).

$$F\Delta\Psi = -RT \ln \frac{[Na]''}{[Na]'} = RT \ln \frac{[Cl]''}{[Cl]'} \tag{2}$$

or, setting

$$\frac{[Na]''}{[Na]'} = \frac{[Cl]'}{[Cl]''} = r$$

$$F \Delta \Psi = -R T \ln r \tag{3}$$

The Donnan PD can be detected by irreversible electrodes. Reversible electrodes which are specific for any of the permeant ions would register only the electrochemical PD of the ion, which is zero in Donnan equilibrium. The difference between reversible and irreversible electrodes will be treated in the last section of this book.

For the simple system above, the dependence of r on the concentration of both the salt (NaCl) and the polyanion can be derived as follows. Since on the right side

$$[Na^+]'' = [Cl^-]'' + nX$$

nX being the equivalent concentration of negative charges of the polyanion, and on the left side

$$[Na^+]' = [Cl^-]' = c'_s,$$

the concentration of the salt on this side, we obtain

$$[Na^+]'' \cdot ([Na^+]'' - nX) = c'^2_s$$

Replacing $[Na^+]''$ by $r \cdot c'_s$ and solving for r we obtain

$$r = \frac{nX + \sqrt{4\,c'^2_s + (nX)^2}}{2\,c'_s} \qquad (4)$$

It is seen that r rises with nX but declines with increasing c'_s. Accordingly a true Donnan potential should be depressed if the charged anionic groups are diminished, e.g., by titration with H-ions, or if the salt concentration is raised.

On the other hand, a true Donnan PD does not depend on active metabolism and therefore is insensitive toward metabolic inhibition. In this respect the Donnan PD may behave differently from the nonequilibrium potentials, which are discussed later. Other permeant electrolytes added to the system will also tend to distribute according to Donnan's law, thereby slightly changing the original distribution, but still reach equilibrium.

From the thermodynamic equilibrium of all permeant ions, it does not always follow that the solvent is also in thermodynamic equilibrium. As the movement of electrolytes toward Donnan equilibrium in our model involves a net transfer of osmolarity into the compartment containing the nonpermeant polyelectrolyte it tends to disturb the equilibrium of the solvent (water). Equilibrium of the complete system therefore requires that a difference in hydrostatic (or osmotic) pressure between the compartments compensates the change of osmolarity, thereby keeping the solvent at equilibrium. The effect of this pressure difference on the solute distribution is negligible. In the above model, this difference in hydrostatic pressure would adjust itself, since the rigidity of walls and barriers prevent appreciable water flow. For a more thorough treatment, the reader is referred to Overbeck (1956).

The Donnan PD between the two sides of a semipermeable membrane, as described in this chapter, is not to be confused with other Donnan PD's which may exist across the two interfaces between the membrane phase and the adjacent solutions. These belong to the phase boundary potentials, which will be discussed in the following section. The terms "Donnan distribution" and "Donnan potential" are often applied incorrectly to systems which as a whole are transient or in a steady state rather than in true equilibrium. There the ion treated as impermeant is only virtually so: it either penetrates very slowly, much slower than the other permeant ions, or it may be maintained at an equal distribution by an active pump. Such "Pseudo-Donnan distributions" will be dealt with later.

1.1.2 Surface Potentials (Phase Boundary Potentials)

These may exist at the phase boundaries between the membrane and the adjacent bulk solutions. They do not usually add to the transmembrane PD in contrast to the transmembrane Donnan PD discussed above, and to other potentials measured between the two adjacent bulk solutions. They do, however, affect the electrical potential profile within the membrane phase and by altering the conductance and permeability properties of the membrane may affect transmembrane potentials indirectly. The complicated subject of such phase boundary potentials and their biological significance is still under extensive study (McLaughlin, 1977). Hence in the present context only the principles of their nature and function can be discussed.

Surface potentials are attributed mainly to "fixed charges" within or attached to the membrane phase, resulting from ionic groups, and to a lesser extent from electric dipolar components of the membrane. Such ionic groups may be there rather permanently or may under special conditions form – or disappear – through the dissociation of weak acidic groups or through association of ions with binding groups, respectively. The contribution of dipoles may be modulated by electric field across the membrane. The relationship between fixed charges and phase boundary potentials can best be illustrated by either one of two simplified borderline models:

1. The first one assumes that the fixed charges are homogeneously distributed over the whole membrane phase, a condition most closely verified by an ion exchange resin.
2. The second one assumes that the charged groups are attached only to the surface of an apolar membrane phase. This model may be best verified by an artificial lipid bilayer membrane.

Though the latter model may come closer to a biological membrane, the quantitative approach is better understood with the former, which may therefore be used here for an illustration.

The surface or phase boundary potentials appearing at an ion exchange membrane can simply be treated in terms of two separate Donnan potentials at each interface (Meyer and Sievers, 1936; Teorell, 1951). The unequal distribution of the permeant ions between each bulk phase and the adjacent region of the membrane phase is associated with an electric potential difference across each boundary. As a consequence, the concentration differences of the permeant ions between the two boundaries of the membrane are different from the corresponding concentration differences between the two bulk phases. For the same reason also the electrical PD across the membrane phase has

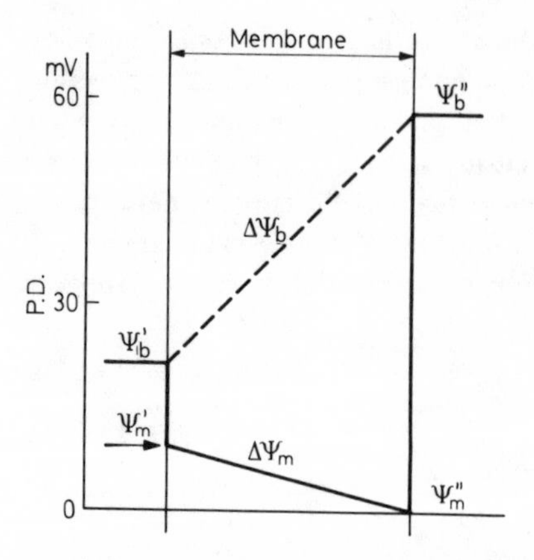

Fig. 2. Presumable profile of electrical potential in a (homogenous) membrane with negative fixed charges. A homogeneous membrane having fixed negatively charged groups at a concentration(nX) of 100 mM is assumed to be in contact with 100 mM HCl on the left, and with 10 mM HCl on the right side. The difference in osmolarity is assumed to be compensated for by a nonpermeant nonelectrolyte. It is assumed that the negative groups of the membranes do not appreciably bind protons under the present conditions. The overall membrane-diffusion potential is assumed to be about 40 mV, right side positive. Due to the unequal Donnan distribution ratios at the two sides ($r'' > r'$), the PD between the two interfaces inside the membrane is different from that between the bulk solutions. Note that the former has an orientation opposite to the latter in this system

to be different from the electrical PD measured between the two bulk phases, as illustrated in Fig. 2. To the extent, however, that quasi-complete Donnan equilibrium is established across the boundaries, as is usually assumed, the electrochemical potential of each permeant ion must be the same on both sides of each boundary. As a consequence, the electrochemical PD of any ion over the membrane phase must be identical with the corresponding difference between the bulk phases. This may be illustrated by a simple system, consisting of two-stirred compartments which contain HCl solutions of different concentrations c'_b and c''_b (Fig. 2), and which are separated by an "ion exchange" membrane with a high density of negative fixed charges. We also assume that an electrical PD is being maintained across the membrane ($\Psi'' - \Psi'$) by a pair of suitable electrodes, so that the overall difference in electrical chemical potential difference of H^+ between the bulk solutions, as indicated by the subscript b, is

$$\Delta\tilde{\mu}_{H^+} = -\left[RT \ln \frac{c''_b}{c'_b} + F\ (\Psi_b'' - \Psi_b')\right] \tag{5}$$

Owing to the Donnan Distribution the concentrations of H-ions within the interfaces on the membrane side would be $r' \cdot c_b'$ and $r'' \cdot c_b''$, r' and r'' standing for the Donnan distribution ratio at the left and right interface, respectively. If $c_b' > c_b''$ then $r'' > r'$, according to the foregoing, so that Δc_m, i.e., the difference in H^+ concentration between the two boundary layers of the membrane phase, which is

$$\Delta c_{\mathrm{m}} = c'_{\mathrm{b}}\, r' - c''_{\mathrm{b}}\, r'' \tag{6}$$

would be smaller than Δc_{b}. There will also be a Donnan PD at each interface, having the value

$$F\,(\Psi'_{\mathrm{m}} - \Psi'_{\mathrm{b}}) = -RT \ln r'$$

and

$$F\,(\Psi''_{\mathrm{b}} - \Psi''_{\mathrm{m}}) = +RT \ln r''$$

respectively. The overall PD within the membrane phase is

$$\Delta\Psi_{\mathrm{m}} = \Psi''_{\mathrm{m}} - \Psi'_{\mathrm{m}} = \frac{RT}{F} \ln \frac{r'}{r''} + \Psi''_{\mathrm{b}} - \Psi'_{\mathrm{b}}, \tag{7}$$

and hence different from $\Delta\Psi_{\mathrm{b}}$. If the ' side is positive, i.e., if $\Psi'_{\mathrm{b}} > \Psi''_{\mathrm{b}}$, $\Delta\Psi_{\mathrm{m}} > \Delta\Psi_{\mathrm{b}}$, the PD between the membrane phase boundaries is greater than that between the bulk solutions. With the inverse orientation of the PD, it will be smaller. In either case, however, the term $RT \ln \frac{r''}{r'}$ will cancel so that the electrochemical PD of the H^+-ions between the two membrane boundaries ($\Delta\tilde{\mu}_{H^+}{}^{\mathrm{m}}$) is

$$\Delta\tilde{\mu}_{H^+}{}^{\mathrm{m}} = RT \ln \frac{c_{\mathrm{b}}''}{c_{\mathrm{b}}''} + F\,\Delta\Psi_{\mathrm{b}} = \Delta\tilde{\mu}_{H^+}{}^{\mathrm{b}} \tag{8}$$

i.e., identical with the electrochemical PD between the bulk phases ($\Delta\tilde{\mu}_{H^+}{}^{\mathrm{b}}$). The same holds true for the corresponding PD's of the Cl-ions. It follows that the electrochemical PD of each ion species across the membrane proper can be obtained from the values measured in the bulk phases even though the partition between chemical and electrical components of the total force is different from that in the membrane phase. To the extent, however, that some permeation process responds to a pure electrical driving force kinetically differently than to an energetically equivalent chemical driving force, and to the extent that the passive permeability of biological membranes is affected by the magnitude of an electrical potential, Donnan effects at the boundaries of membrane phases may occasionally be of importance. Since biological membranes do not appear to have a high density of fixed charges, the Donnan effects at the phase boundary are presumably of much smaller magnitude here than in true ion exchange membranes.

The analogous approach to the second borderline model, that of the charge on a nonpolar membrane phase, is in principle similar but in its detail more complicated. This is mainly due to the fact that the boundary across which the main potential jump is located need not be identical with the interface of the membrane phase, from which it may be removed some ten Å (Debye distance). For this reason, this PD is less sharply defined and though a Donnan-like distribution should in principle be established between the bulk solution and a layer in the immediate vicinity of the membrane phase, a precise description is not possible in these terms. The various approaches and theories have been discussed by Bockris and Reddy (1973), and more recently by McLaughlin (1977) to which the interested reader may be referred for further details and references. At the present juncture, it may suffice to say that the conclusion we derived from the first model, that of a double Donnan distribution at the interface boundaries (see Fig. 2), in principle holds also for the second one, namely: the free energy of an ion is hardly changed by passing through a phase boundary, even though the profiles of the electrical and of the chemical potential within the membrane may be quite different from, even inverted to, that derived from the difference between the two bulk solutions. It seems, however, that the permeability and conductance properties of the membrane phase are influenced more strongly by the electrical potential profile than by the ion concentrations. Hence the variations of the phase boundary potential may profoundly modulate the permeability of biological membranes, e.g., by altering the state of pores and channels, and may occasionally give rise to rectification and other alterations of the current-voltage relationship (McLaughlin, 1977; Dilger et al., 1979).

As already mentioned, the phase boundary potentials are, in contrast to "transmembrane" potentials, not usually measurable simply by electrodes between the two adjacent solutions. In order to obtain information about their magnitude, certain channel-forming ionophores are used as "probes" ("molecular voltmeter") (Mueller and Finkelstein, 1972; Finkelstein and Andersen, 1980). The mechanism by which they function is complicated and not completely elucidated and hence will not be discussed in the present context.

1.2 Membrane Diffusion Potentials

1.2.1 General

Whenever an electrolyte diffuses down its concentration gradient, a PD may be measured between the two regions of lower and higher concentration, except if both cationic and anionic species have the same mobility. Orientation and magnitude of this "diffusion PD" depends on the difference in mobilities between the cationic and anionic species of electrolyte. Normally the potential of the more dilute region will have the same sign as the faster ion, and vice versa, and the PD will be greater, the greater the difference between the cationic and anionic mobilities.

Diffusion PD's in free solution are usually very transient and easily disturbed by convectional flows of the solution, and therefore of little practical significance. Diffusion PD's may, however, be more stable if the two solutions containing the electrolytes at different concentrations are in contact with each other in a way that allows some diffusion but hinders convectional mixing. Such a contact may be direct (liquid junction) or through a separating "permselective" membrane, i.e., a membrane which is permeable to different ion species to a different extent. The diffusion PD at liquid junctions is a possible source of error whenever an electrical PD is measured by electrodes using "salt bridges", as will be discussed in a later section (3.2). The membrane diffusion PD, on the other hand, is more common in biological systems and will therefore be treated more extensively in the present context. The term "membrane diffusion" PD is not identical with the term "membrane" PD, which is used for a membrane that is permeable exclusively to either cations or anions. In that case the maintenance of electroneutrality will (in the absence of an outside EMF) prevent net diffusion of electrolyte, whereas the diffusion PD has its maximum value. The membrane PD can be considered a borderline equilibrium PD.

1.2.2 Systems with One Electrolyte

To illustrate *the principles of a membrane diffusion PD* we use the same model as before but assume that in each of the two aqueous compartments there is HCl, but at a concentration higher on the left side than on the right one (Fig. 3). The membrane is assumed

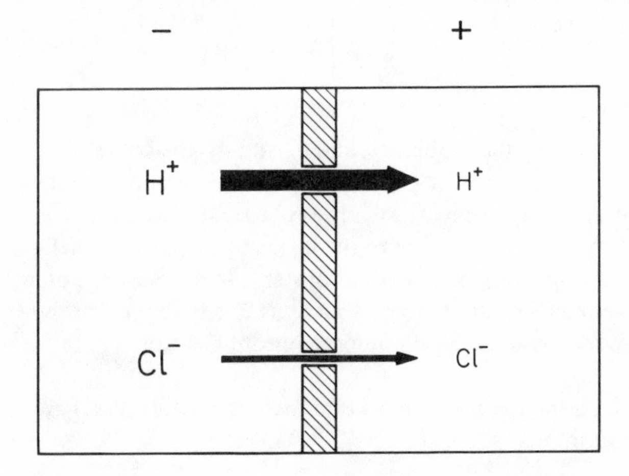

Fig. 3. Membrane-diffusion potential. HCl is present at different concentrations in two compartments, which are separated by a membrane (*shaded area*). As indicated by the thicknesses of the arrows, H-ions are more permeant than Cl-ions. As a consequence, the right side membrane appears to be electrically positive relative to the left side. For details see text

to be permeable to each species, but more so to H^+ than to Cl^-. Such membranes, which are called "permselective", exist in nature or can be artificially prepared, e.g., by the incorporation of anionic substances in the membrane matrix. While both ion species move toward equilibrium distribution, the H-ions tend to "run ahead" through the membrane but are retained by the electrostatic forces caused by a "microscopic" separation of H^+ from Cl^- in the membrane region, rendering the right-hand solution positive relative to the left-hand solution. This apparent violation of the principle of electroneutrality is, as stated, only microscopic, i.e., it is usually not detectable by chemical means. Only if the membrane area is very large relative to the size of the separated compartments may deviations from electroneutrality become substantial. Since the maintenance of electroneutrality requires that both H^+- and Cl^--ions penetrate the membrane at the same speed, the electrochemical potential difference of the faster H^+-ions will at all times be smaller than that of the slower Cl^--ions. Hence a transient "Donnan"-like state may obtain during time periods which are short enough for the Cl-ions to be treated as the virtually impermeant species, and the H^+-ions as being in quasi-equilibrium. In contrast to a true Donnan equilibrium, however, the quasi-equilibrium distribution of H^+ will slowly but permanently drift as Cl^- passes from the left to the right compartment.

To express the membrane diffusion PD in this model as a function of the concentration of the salt and of the individual ion mobilities (u_i), we start from *the general flux equations* for ions as devised by Nernst (1888).

$$J_+ = -u_+ \cdot c_+ \left[RT \frac{d(\ln c_+)}{dx} + \frac{F\,d\Psi}{dx} \right]^{1)} \tag{9a}$$

$$J_- = -u_- \cdot c_- \left[RT \frac{d(\ln c_-)}{dx} - \frac{F\,d\Psi}{dx} \right]^{1)} \tag{9b}$$

J_+ and J_- are the (net) flows, u_+ and u_- the mobilities, and c_+ and c_- the concentrations of cation and anion, respectively. x is the distance within the membrane from the left interface in the direction perpendicular to it. J_i will always be treated as positive if its direction is from left to right, or from the outside to the inside of a cell or organelle. R and T have the conventional meaning, F is the Faraday constant, Ψ the electric potential. For better understanding, we shall first apply these equations to a binary electrolyte by assuming that in our above model besides HCl no other permeant electrolytes are present.

Since the maintenance of electroneutrality requires that the two ion net fluxes are equivalent, which for a binary salt means that

$$J_+ = J_-$$

we can combine Eqs. (9a and b) to solve for d Ψ, the differential of the electrical potential difference:

$$F\,d\Psi = -\frac{u_+ - u_-}{u_+ + u_-} RT\,d(\ln c) \tag{10}$$

---

1 The electrical term should be multiplied by z_i, the valency of the ion concerned, but for the monovalent cation and anion of the present electrolyte z_i can be replaced by +1 and −1 respectively c in the logarithmic form should, in contrast to the factorized c, be the chemical activity (a_i) of the ion concerned. The activity coefficient has been neglected assuming that it does not change appreciably with the distance

Integration of this equation over the whole thickness of the membrane (δ) is simple under the following conditions:

1. The membrane is a homogeneous phase, in which the ion mobilities (u) as well as the activity coefficients are the same for any value of x.
2. The membrane has no fixed charges so that the concentrations of anion and cation are the same at any cross-section of the membrane.
3. The membrane is "unstirred", i.e., the migration of ions through the membrane phase is by free diffusion only.

These assumptions may be unrealistic but helpful to illustrate the basic relationships involved. The integration leads to the following well-known equations

$$F \int_0^\delta d\Psi = -\frac{u_+ - u_-}{u_+ + u_-} RT \int_0^\delta d(\ln c)$$

$$F \Delta\Psi = -\frac{u_+ - u_-}{u_+ + u_-} RT \ln\left(\frac{c''}{c'}\right) \tag{11}$$

c' and c'' strictly referring to the concentrations in the two boundary regions of the membrane. Provided that there is quasi-equilibrium between each boundary region and the adjacent bulk solution, the concentrations c' and c'' are equal to the corresponding concentrations in the adjacent bulk solutions times a partition coefficient. For a homogenous membrane, the partition coefficient should be the same on both sides and hence cancel in the above equation, so that the ratio $\frac{c'}{c''}$ also stands for the corresponding ratio of the concentrations (or activities) in the bulk solutions.

We arrive at a similar equation if we apply the *principles of irreversible thermodynamics* (Kedem and Essig, 1965). We start with the following differential equations:

$$r_{H^+} J_{H^+} = -RT \frac{d(\ln c)}{dx} - F \frac{d\Psi}{dx} \tag{12a}$$

$$r_{Cl^-} J_{Cl^-} = -RT \frac{d(\ln c)}{dx} + F \frac{d\Psi}{dx} \tag{12b}$$

r_H and r_{Cl} being the specific resistance coefficients for H^+ and Cl^-, respectively, at the distance x within the membrane phase. Assuming that the membrane neither a accumulates nor releases HCl, we can integrate each equation over the whole thickness (δ) of the membrane, i.e., between $x = 0$ and $x = \delta$:

$$J_{H^+} \cdot R_{H^+} = -RT\,\Delta\ln c - F\Delta\Psi \tag{12c}$$

$$J_{Cl^-} \cdot R_{Cl^-} = -RT\,\Delta\ln c + F\Delta\Psi \tag{12d}$$

We now use R_{H^+} and R_{Cl^-}, the *integral resistance* coefficients for cation and anion, respectively, which are defined as follows:

$$R_{H^+} = \int_0^\delta r_H \cdot dx,$$

$$R_{Cl^-} = \int_0^\delta r_{Cl} \cdot dx$$

and must be determined empirically.

Since according to the principle of electroneutrality

$$J_{H^+} = J_{Cl^-}$$

we can solve the equations for $F\Delta\Psi$

$$F\Delta\Psi = -\frac{R_H - R_{Cl}}{R_H + R_{Cl}} RT \ln\left(\frac{c''}{c'}\right) \tag{13}$$

For a homogenous membrane Eq. (13) becomes identical with Eq. (11) since

$$r_i = \frac{1}{u_i c} \quad \text{and} \quad R_i = \int_0^\delta \frac{dx}{u_i c}$$

and as in such a membrane each mobility should be the same for any x, we obtain

$$R_{H^+} = \frac{1}{u_+} \int_0^\delta \frac{dx}{c} \quad \text{and} \quad R_{Cl^-} = \frac{1}{u_-} \int_0^\delta \frac{dx}{c}$$

which in combination with Eq. (13) lead directly to Eq. (11).

In the absence of any coupling the integral resistance is equal to the inverse phenomenological leakage coefficient: $R_i = \frac{1}{L^u_{ii}}$, hence

$$J_{H^+} = L^u_{H^+} (-RT \Delta \ln c_{H^+} - F\Delta\Psi) \tag{14a}$$

$$J_{Cl^-} = L^u_{Cl^-} (-RT \Delta \ln c_{Cl^-} + F\Delta\Psi) \tag{14b}$$

$$F\Delta\Psi = \frac{L^u_{H^+} \cdot RT \Delta \ln c_{H^+} - L^u_{Cl^-} \cdot RT \Delta \ln c_{Cl^-}}{L^u_{H^+} + L^u_{Cl^-}} \tag{14c}$$

This equation can immediately be converted into that of an electrical equivalent circuit, expressing all terms by their electrical correlates. Since the specific conductance $g_i = F L^u_i$, and as the terms

$$-\frac{RT}{F} \Delta \ln c_{H^+} \quad \text{and} \quad -\frac{RT}{F} \Delta \ln c_{Cl^-}$$

are usually represented by their "Nernst potentials", E_{H^+} and E_{Cl^-}, respectively, i.e., the electrical PD's with which the distribution of the respective ion would be in equilibrium, we obtain

$$\Delta\Psi_m = \frac{g_{H^+} \cdot E_{H^+} + g_{Cl^-} \cdot E_{Cl^-}}{g_{H^+} + g_{Cl^-}} \tag{15}$$

As to the *reaction order of ion diffusion* in the above systems, we shall not enter an extensive treatment of the kinetics of ion transport but try to illustrate the basic principles with the help of an example, namely the passive movement of HCl across the barrier between the well-stirred compartments of the system described above. For simplicity reasons we again disregard osmotic water flow by assuming rigid walls and a rigid membrane. We also assume that HCl is the only electrolyte present in the two compartments. We shall consider the two following borderline cases: (1) H^+ and Cl^- penetrate the barrier only as a neutral complex, e.g., either as undissociated HCl or as cotransport with a neutral carrier; (2) H^+ and Cl^- penetrate the barrier separately, each diffusing down its own electrochemical activity gradient, without interacting with each other. Complications due to unstirred layers should be insignificant if in either case the penetration through the barrier is slow as compared to the diffusion through the unstirred layers of the bulk solutions. To start with the first case, we postulate that K_{HCl} is the dissocation of HCl, α and β are the coefficients of the rates at which HCl enters and leaves the membrane phase, respectively, at either side, and p is the rate coefficient at which the HCl complex moves within the membrane phase. We obtain in the steady state for the overall rate at which both H^+ and Cl^- penetrate the membrane:

$$J_{HCl} = -P_{HCl} \cdot \Delta c_{HCl}^2 \tag{16}$$

$$P_{HCl} = \frac{p \cdot \alpha}{(2p + \beta) K_{HCl}},$$

Δc_{HCl} being the difference in concentration of HCl between the two compartments. We see that in this case the penetration rate is a *second order process.*

If H^+ and Cl^- were to pass the membrane exclusively by electrically silent cotransport via a common carrier, the process would still be of the second order, though the saturability of the transport system introduces some complications.

In the second case, in which H^+ and Cl^- pass the membrane separately, the general diffusion equations of the positive and negative ion species, respectively, within the membrane phase hold [Eqs. (9 a, b)].

For a single electrolyté, $c_+ = c_- = c$, we can replace according to Eq. (10)

$$F \frac{d\Psi}{dx} \text{ by } -\frac{u_+ - u_-}{u_+ + u_-} RT \frac{dc}{c \cdot dx} \quad \text{in each equation,}$$

and obtain

$$J_+ = J_- = \frac{2u_+ \cdot u_-}{u_+ + u_-} RT \frac{dc}{dx}$$

If we set $\frac{2u_+ \cdot u_-}{u_+ + u_-} RT = D_{HCl}$, to form an overall diffusion coefficient, the equation becomes identical with Fick's diffusion equation (I)

$$J_{HCl} = -D_{HCl} \cdot \frac{dc}{dx}$$

or

$$J_{HCl} = -P_{HCl} \cdot \Delta c \tag{17}$$

We see that in this case the diffusion of HCl is a *first order process*; in other words, HCl moves as if it were a single species.

(This may not be as clear if we consider the membrane a nonhomogeneous phase with a high, nonelectric energy barrier. In that case the individual ion fluxes are, as will be discussed later [see Eq. (22a and b)]:

$$J_+ = P_+ \; (c' \xi^{\frac{1}{2}} - c'' \xi^{-\frac{1}{2}})$$

$$J_- = P_- \; (c' \xi^{-\frac{1}{2}} - c'' \xi^{\frac{1}{2}})$$

Setting $J_+ = J_-$ and rearranging, we get the salt diffusion rate

$$J = \xi^{-\frac{1}{2}} \; \frac{P_+ \, P_-}{P_+ c' + P_- c''} \; (c'^2 - c''^2)$$

or, solving for $\xi^{-\frac{1}{2}}$

$$J = \frac{P_+ \cdot P_- \cdot (c' + c'')}{\sqrt{(P_+ c' + P_- c'') \, (P_+ c'' + P_- c')}} \cdot (c' - c'')$$

The overall permeability coefficient is dependent on the potential. The order of reaction, however, is still unity or less, as can be seen if we replace $\xi^{-\frac{1}{2}}$ by the corresponding expression derived from Eq. (20)).

1.2.3 Systems with Several Electrolytes

If the concentrations of the permeant cations and anions are not equal in each chamber, or if more than one salt of permeant ions is present, the quantitative treatment becomes very complicated, mainly because, for obvious reasons, the concentration gradient of a single ion species and the electrical potential gradient across the membrane cannot both be linear. A complete treatment of such systems was carried out by Planck (1890) and more recently by Schlögl (1964), but the resulting equation is too involved to be of practical use for biological systems. Two major attempts have been made to simplify it: the best known is that by Goldman (1944), who treated the electrical potential gradient as linear, so that the concentration gradient must be nonlinear. The other one is by Henderson (1907), who treated the concentration gradient as linear so that the electrical potential gradient cannot be linear. Each of these simplified approaches leads to an equation which is practically useful under certain conditions. For a more thorough treatment the reader is referred to the treatise by Schlögl (1964).

1.2.3.1 Approximation Approaches

For a better comparison, we shall in the following derive both the Goldman equation and the Henderson equation for the simplified model described above, except that it is no longer restricted to a single binary electrolyte. We shall, however, keep the system simple by dealing only with the same two permeant ion species as above, though these, in contrast to the previous system, need no longer have identical concentrations in each compartment, as other, nonpermeant ion species may be present as well, i.e., $c'_+ \neq c'_-$, $c''_+ \neq c''_-$. Furthermore, we retain the assumption of a homogeneous membrane, in which u_+ and u_- do not change with x.

In order to integrate these equations we may introduce either one of the already-mentioned simplifying assumptions, i.e., of a linear gradient of either the electrical potential (Goldman) or of the solute concentration (Henderson). Let us start with *Goldman's approach*, i.e., by assuming a constant electrical field. Hence, we set in both equations (9a, 9b)

$$\frac{d\Psi}{dx} = \frac{\Delta\Psi}{\delta}$$

i.e., the gradient of the electrical potential at any place within the membrane is identical with the total difference in electrical potential divided by the thickness of the membrane phase (δ). After separating the two variables (x and c) we obtain:

$$\left(J_+ + u_+ \frac{F\Delta\Psi}{\delta} c_+\right) dx = -u_+ RT\, dc_+$$

and

$$\left(J_- - u_- \frac{F\Delta\Psi}{\delta} c_-\right) dx = -u_- RT\, dc_-$$

Integrating these equations over the thickness of the membrane (δ) we obtain for the cation

$$-\frac{F\Delta\Psi}{RT} = \ln\left(\frac{\delta J_+ + u_+ F\Delta\Psi\, c''_+}{\delta J_+ + u_+ F\Delta\Psi\, c'_+}\right) \qquad (18a)$$

and for the anion

$$-\frac{F\Delta\Psi}{RT} = -\ln\left(\frac{\delta J_- - u_- F\Delta\Psi c''_-}{\delta J_- - u_- F\Delta\Psi c'_-}\right) \tag{18b}$$

As in a homogeneous membrane

$$\frac{u_+}{\delta} = \frac{P_+}{RT}$$

we obtain for the ion flows by solving Eqs. 18a and 18b

$$J_+ = P_+ \frac{F\Delta\Psi}{RT} \frac{c'_+ e^{-\frac{F\Delta\Psi}{RT}} - c''_+}{1 - e^{-\frac{F\Delta\Psi}{RT}}} \tag{19a}$$

$$J_- = P_- \frac{F\Delta\Psi}{RT} \frac{c'_- - c''_- e^{-\frac{F\Delta\Psi}{RT}}}{1 - e^{-\frac{F\Delta\Psi}{RT}}} \tag{19b}$$

$e^{-\frac{F\Delta\Psi}{RT}}$ is often replaced by ξ, the "electrochemical activity coefficient."

In these equations the reference PD is arbitrarily set equal to zero on the "transside" of the system. For some cases it may be more useful to place this reference at the middle of the membrane ($\frac{\delta}{2}$). The above equations would then read:

$$J_+ = P_+ \frac{F\Delta\Psi}{RT} \frac{c'_+ \xi^{\frac{1}{2}} - c''_+ \xi^{-\frac{1}{2}}}{\xi^{-\frac{1}{2}} - \xi^{\frac{1}{2}}} \tag{19c}$$

$$J_- = P_- \frac{F\Delta\Psi}{RT} \frac{c'_- \xi^{-\frac{1}{2}} - c''_- \xi^{\frac{1}{2}}}{\xi^{-\frac{1}{2}} - \xi^{\frac{1}{2}}} \tag{19d}$$

$\xi^{-\frac{1}{2}} - \xi^{\frac{1}{2}}$ can also be written $2 \sinh \frac{F\Delta\Psi}{2\,RT}$.

As electroneutrality requires that $J_+ = J_-$ we can solve for the electrical potential:

$$F\Delta\Psi = -RT \ln \frac{P_+ c''_+ + P_- c'_-}{P_+ c'_+ + P_- c''_-}$$

or

$$\xi = \frac{P_+c''_+ + P_-c'_-}{P_+c'_+ + P_-c''_-} \tag{20}$$

The question arises whether for a single electrolyte ($c'_+ = c'_-, c''_+ = c''_-$) the Goldman equation and the previously derived Nernst equation [Eq. (11)] give approximately the same result, if we use the same mobility units in both cases. The answer is, however, not necessarily yes. The disagreement between the two approaches is the smaller, the greater the difference between the mobilities of cation and anion, respectively. If one of the mobilities becomes zero, so that we obtain a true membrane potential without any diffusion component, both equations give the same answer. If the permeability coefficients come closer to each other, the discrepancy becomes greater, but may still be small if the difference in concentrations between the two compartments is also small, in other words, if the system is close to equilibrium. The discrepancy is due to the simplifying assumption of a constant electrical field in the Goldman approach.

Let us now investigate whether the corresponding equation derived on the basis of *Henderson's approach,*, i.e., by assuming a constant concentration gradient, gives a better agreement. For this purpose we assume for each ion species a constant concentration gradient, i.e.,

$$\frac{dc}{dx} = \frac{\Delta c}{\delta}$$

so that we can replace c'' by $c' + \frac{\Delta c}{\delta} x$.

Starting again from Eqs. (9a and b), we eventually obtain by integration between $x = 0$ and $x = \delta$ the equation

$$F\Delta\Psi = -\frac{u_+\Delta c_+ - u_-\Delta c_-}{u_+\Delta c_+ + u_-\Delta c_-} RT \ln \frac{u_+c''_+ + u_-c''_-}{u_+c'_+ + u_-c'_-} \tag{21}$$

If only one (binary) salt is present, i.e., if $c_+ = c_-$ in both compartments, the equation reduces to

$$F\Delta\Psi = -\frac{u_+ - u_-}{u_+ + u_+} RT \ln \frac{c''_i}{c'_i}$$

which is identical with Eq. (11). Equation (21) can easily be expanded to include more than one salt. We may conclude that the assumption of a linear concentration gradient for both ions of a single binary salt is under these conditions verified, so that for such a simple system the Henderson approach is more appropriate than the Goldman approach, which, as we have seen, introduces considerable deviations. For biological systems, however, the Goldman approach is usually preferred to the Henderson approach for the following reasons: Firstly, in systems with various permeant ion species on both sides of the membrane, the Goldman equation has been shown to give more realistic results than does the Henderson equation, especially if the total ion concentration on the one side is similar to that on the other side of the barrier, as is usually the case in biological systems. Secondly, for biological membranes, which are not homogeneous but contain a lipid core which is considered a barrier with an especially high activation energy for ions; the assumption of a constant concentration gradient is unrealistic. It seems that in biological systems the relation between electrical PD and the permeability behavior of ions is better described in the Goldman equation, whereas the Henderson equation gives the better result in a diffusional system with only one electrolyte present.

1.2.3.2 The Statistical Approach

This may be more appropriate than the above approaches if the (activation) energy required to overcome the lipid barrier is much greater than the electrical potential difference across the membrane, as is likely to be true for biological membranes. The rate of movement of a single ion should here not depend appreciably on the electrical field but rather on how many particles have a free energy just high enough to overcome this barrier. The probability (P_i^*) that any solute particle meets this requirement can be estimated from the Boltzmann equation in which it is an exponential function of ΔE^{+}, the avtivation energy required to overcome the energy barrier $\emptyset$. (Fig. 4)

$$P_i^* = A\, e^{-\frac{\Delta E^{\ddagger}}{RT}}$$

A is a proportionality constant. We now assume that whenever a particle has such a level, it overcomes the barrier at a high speed, independent of the electrical field. For such a system the complicated flow equation may often be simplified as follows.

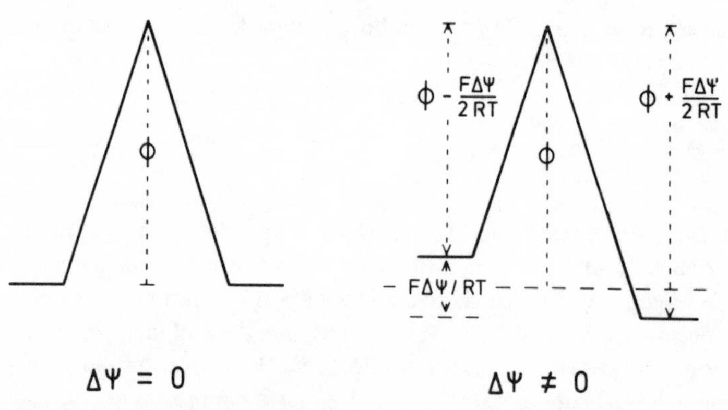

Fig. 4. Osmotic barrier. The ions penetrating the membrane are supposed to "jump" over an osmotic barrier, which is large as compared to the electrical PD. For simplicity reasons it is assumed that the peak of the barrier is in the middle between the two boundaries of the membrane. In this model the movement of cations and anions across the membrane is approximately represented by Eqs. (22a and b)

$$J_+ = P_+^* \left(c'_+ \xi^{\frac{1}{2}} - c''_+ \xi^{-\frac{1}{2}}\right) \tag{22a}$$

$$J_- = P_-^* \left(c'_- \xi^{-\frac{1}{2}} - c''_- \xi^{\frac{1}{2}}\right) \tag{22b}$$

Obviously the flow is taken here to be proportional to the difference between the electrochemical activities relative to the innermost layer of the membrane. It is implied that the activation barrier is a single peak which is located in the middle between the two membrane faces. This assumption may not always be correct but deviations are not likely to cause major alterations of the above equation (Parlin and Eyring, 1954; Hall et al., 1973).

Equations (22a and b) come at least qualitatively closer to reality for many bilayer systems than do the corresponding Goldman equations [Eq. (19)] (Läuger and Neumcke, 1973). This can be illustrated by comparing the current voltage relationship of such membranes with that predicted by our equations for equal ion concentrations on both sides of the membrane. Equation (19) under such conditions ($c' = c'' = c$) reduces to

$$J_+ = -P_+ c \frac{F\Delta\Psi}{RT} \quad \text{and} \quad J_- = P_- c \frac{F\Delta\Psi}{RT}$$

so that the current produced by an outside PD becomes

$$I = F(J_+ - J_-) = -(P_+ + P_-)c \frac{F^2\Delta\Psi}{RT}$$

The system is "ohmic". Equation (22), however, under the same conditions reduces to

$$J_+ = P_+^* c \left(\xi^{\frac{1}{2}} - \xi^{-\frac{1}{2}}\right) \quad \text{and} \quad J_- = P_-^* c \left(\xi^{-\frac{1}{2}} - \xi^{\frac{1}{2}}\right)$$

and hence the current

$$I = F(J_+ - J_-) = (P_+^* + P_-^*)\, cF \left(\xi^{\frac{1}{2}} - \xi^{-\frac{1}{2}}\right)$$

$$= (P_+^* + P_-^*)\, c\, 2F \sinh \frac{F\Delta\Psi}{RT}$$

is no longer a linear but rather a sigmoid function of the PD (Fig. 5). This kind of function is at least qualitatively in better agreement with the current voltage relationship of artificial lipid bilayers as well of some, though perhaps not all, biological membranes.

1.2.3.3 The Goldman-Hodgkin-Katz Equation

In the more general form as it is in practical use now, the equation was formulated on the basis of Goldman's derivation by Hodgkin (1957) and describes the relation between

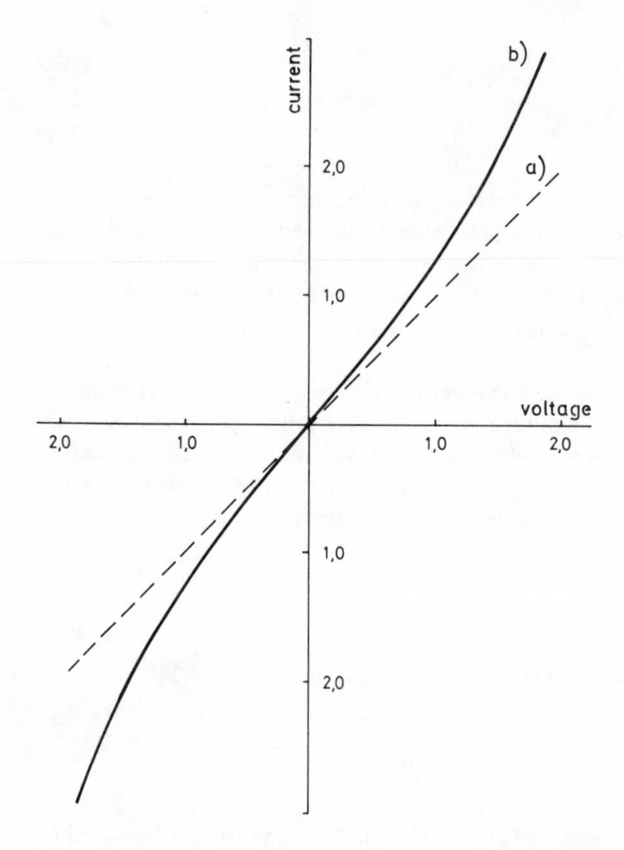

Fig. 5. Current-voltage relationship. *a* As predicted from the Goldman equation showing "ohmic" behavior. *b* The relationship predicted is the ions' jump over an energy barrier. The curve is sigmoid. For details see text

(transient) electrical potential on the one hand, and the concentrations and permeancies of the ions on the other:

$$\Delta\Psi = -\frac{RT}{zF} \ln \frac{P_{Na}\,[Na^+]'' + P_K\,[K^+]'' + P_{Cl}\,[Cl^-]' + \ldots}{P_{Na}\,[Na^+]' + P_K\,[K^+]' + P_{Cl}\,[Cl^-]'' + \ldots} \tag{23}$$

As the same equation results from either the Goldman Eqs. (19a and b) or the statistic Eqs. (22a and b), there is in this context no reason to differentiate between P_i (permeability coefficient) and P_i^* (probability coefficient).

Under special conditions certain terms may be zero or so small that they can be omitted, which will greatly simplify the equation. Also, terms referring to ions which have reached complete equilibrium distribution can be omitted, since they cancel. This applies to passive ions in the presence of ion pumps in the steady state, as we shall see later. A typical example is a proton pump in the presence of highly permeant K^+, which in the steady state maintains H-ions at disequilibrium whereas the passive K-ions reach true equilibrium with the membrane potential maintained and therefore drop out of the above equation. Errors may, however, occur if the steady state is not yet complete, because even close to equilibrium passive ions with high permeability coefficients may make much difference. At any event, therefore, as long as the system is transient, it is dangerous to omit any term referring to an ion permeating at an appreciable rate.

Of practical interest is a borderline case of the above equation, which obtains if only *cations* penetrate the membrane at an appreciable rate. Now the terms with the anions can be neglected, at least for transient states. A simple example would be a two-compartment system with a cation permselective barrier which has an NaCl solution on the right side and a KCl solution of equal concentration on the left side. We assume that the permeability coefficient for K^+ is greater than that for Na^+ whereas that of Cl^- is negligible. Obviously, each cation species can leaves its compartment only by a one-to-one exchange with the other, but the more rapid K^+ ions try to "run ahead", thereby being restrained by electrostatic forces. Owing to a microscopic separation of charges the membrane will be positive on the right and negative on the left side. As in previous cases, this potential will be higher the greater the difference in permeability of the membrane for the two cations concerned. In the initial stage the Goldman-Hodgkin-Katz equation [Eq. (23)], provided the concentration of NaCl and KCl are equal on both sides of the membrane, reduces to

$$\Delta\Psi = \frac{RT}{zF} \ln \left(\frac{P_K}{P_{Na}}\right). \tag{24}$$

During the subsequent process of cation exchange, the distribution of the faster ion (K^+) will at any time be closer to equilibrium with the electrical PD than that of the slower ions (Na^+).

The system becomes more complicated if the permeability for Cl^- ions is very high as compared to those of Na^+ and K^+ which are lower but, as above, unequal. Since the electrical potential is now predominantly determined by the chloride distribution, it would be largely suppressed under the present conditions. Owing to the high mobility, the chloride distribution is almost in equilibrium with the electrical potential, but not quite so because chloride still migrates together with K^+ to the other compartment. Consequently, there is some net transfer of KCl across the membrane so that the exchange between K^+ and Na^+ is no longer 1:1. This transfer will proceed until the chloride distribution ratio has reached equilibrium with the electrical potential and from then on the exchange of Na^+ and K^+ will be precisely 1:1. But soon afterwards Cl^- will start to move backward, together with a small excess of Na^+ flow over the opposite K^+ flow because in the final equilibrium chloride should be equally distributed between both sides of the membrane. We may learn from this example that in transient states the chloride term cannot be omitted from the equation (23), even though its distribution is most of the time very close to its equilibrium with the PD except for the short time period mentioned in which the potential is exclusively determined by the Na^+ and K^+ terms. Most of the time, however, the chloride term does contribute to the PD and it is difficult to decide when this contribution becomes small enough to be neglected. In summary, the simplified equation presented in the previous paragraph can only be applied to the extent that the anion movement is truly negligible.

1.2.4 Stabilization of Membrane-Diffusion Potential Differences

As has been pointed out, the previously described systems with membrane-diffusion PD's are transient and necessarily drift continuously toward true equilibrium. They may however be transformed into *steady state systems* by introducing ion pumps to maintain the constant nonequilibrium distribution of one or more ion species. Though the ion pumps and their direct effects on membrane PD's will be treated in more detail later, we shall briefly consider them in the present context, but only with respect to their stabilizing effect on membrane-diffusion potentials. Accordingly we may extend the previous model by an Na^+ "pump", which by actively transporting Na^+ from the right to the left compartment continuously restores the Na-ions to the left compartment at the rate at which they escape. In the steady state this membrane would behave as if it were (virtually) impermeable to Na-ions, whereas the distribution of the permeant Cl-ions has attained complete equilibrium. At face value such a system closely resembles a Donnan system in which the unequally distributed Na-ions assume the function of a nonpermeant (polyionic) cation species. This is because all the permeant ion species present, let us say K^+ and Cl^-, would in the steady state be distributed as in a Donnan system:

$$\frac{[K^+]''}{[K^+]'} = \frac{[Cl^-]'}{[Cl^-]''}$$

and the electrical PD would be

$$F\Delta\Psi = -RT\,\Delta\ln[K^+] = RT\,\Delta\ln[Cl^-]$$

Although the term Donnan distribution has often been applied to such a system, it differs fundamentally from a true Donnan system in several respects, so that the term *pseudo-Donnan system* would be more appropriate. For instance, in the latter system the ion distribution depends on metabolic energy supply because if the pump stops working, the Donnan-like distribution of K- and Cl-ions will eventually collapse. But also with an intact pump a pseudo-Donnan system might behave differently from a true Donnan system, for instance if the concentration of the passive ion is changed. Whereas the distribution of a truly impermeant ion may not be affected by such a change, that of an actively transported ion probably will be. Increasing the concentration of KCl in our model system should in both cases decrease the PD but probably at the same time stimulate Na^+ net transport until a new steady state is reached in which the difference in concentration of the Na-ions between the compartments would be greater than before. As a result the reduction in electrical potential, as caused by the addition of KCl, would be smaller than under the same circumstances in the true Donnan system. Also a decrease in the KCl concentration would affect a true Donnan system differently from our pseudo-Donnan system. It would in either system tend to increase the electrical PD, but much more strongly in a true Donnan system, i.e., with true impermeability of the membrane to Na^+, than in the pseudo-Donnan system. Theoretically it would tend toward infinity as the concentration of KCl approached zero. In our pseudo-Donnan system, however, owing to the limited power of the pump, the difference in Na^+ concentration between the two compartments would have to decrease under these circumstances. The rise in electrical potential would hence be smaller than in a true Donnan system as it could not exceed the ceiling determined by the power of the Na^+ pump. These examples may illustrate that a membrane potential in the steady state, in spite of some similarities, behaves in important respects quite differently from a true Donnan system.

1.3 Electrogenic Pump Potentials

1.3.1 General

The pumps of electrolyte ions are divided into neutral or *electrically silent pumps* and *electrogenic pumps*. The former may act by pumping ions of equal charge and equal sign in opposite directions, or electrically equivalent ions of opposite charge in the same direction, at such stoichiometry that no net electric charge is translocated. Typical examples are the NaCl pump in the gall bladder and the K^+-H^+-exchange pump in the stomach, respectively. The electrogenic pump, in contrast, effects a net movement of electric charge across the membrane, thereby acting as a generator of electric energy. It may act by transporting a single ion species only, or by forcibly exchanging two different ion species of the same sign but different charge at such stoichiometry that a net transfer of charge results. Typical examples are the electrogenic proton pumps in mitochondria and bacteria, and the Na^+-K^+-pump in cellular plasma membranes, respectively (1.3.2.1).

Both electrically silent and electrogenic pumps may give rise to an electric potential. The electrically silent pump can do so only indirectly, by keeping the pumped ions at a disequilibrium distribution. Whenever the passive mobilities of the ions are different from each other a membrane diffusion PD will result according to the Goldman-Hodgkin-Katz equation [Eq. (23)].

By contrast, the electrogenic pump may affect the electrical PD directly by a contribution of its own, in addition to its indirect effect on a membrane diffusion PD through the redistribution of ions. Only this latter part can be described by the unmodified Goldman-Hodgkin-Katz equation [Eq. (23)]. Appropriate equations that take into account this direct contribution of the pump to the PD will be derived in a later section (1.3.2).

A pump working either at constant rate or at constant driving affinity is energetically limited by a "ceiling", i.e., by the highest electrochemical PD of the transported solute that the pump can achieve and maintain at a given steady state, under constant conditions. The ceiling is at its maximum at static head. But also in other steady state conditions, for instance, at stationary flow, there may be a similar, though somewhat lower, ceiling, as long as the net flow rate is fixed. The ceiling can be defined most clearly in terms of Thermodynamics of Irreversible Processes (TIP), which permits a simple separation of chemical and electric potentials, as will be shown later (1.3.2). Before deriving the detailed equations in these terms, a crude, qualitative approach may be helpful for an intuitive understanding of the basic principles. For this purpose we shall now look at two arbitrary models of salt transport: an electrically silent NaCl pump and an electrogenic Na pump, respectively, which in a membrane assumed to be permeable to Na and Cl only, should both lead to concentrative transfer of NaCl. Whereas the first of these models may come close to NaCl transport in the gall bladder (Diamond, 1962), the second one, at least in the simple form considered here, is of doubtful reality. But these models are chosen here for their simplicity, regardless of their biological significance.

For an *electrically silent NaCl pump*, the ceiling is – as will be formally derived later – the sum of the electrochemical PD's of Na^+ and Cl^-. As the electric PD's implicit in these terms cancel here, the ceiling is completely defined by the sum of the chemical PD's:

$$RT\,\Delta\ln\,[Na^+] \;+\; RT\,\Delta\ln\,[Cl^-] \;=\; \text{ceiling} \tag{25a}$$

It thus contains osmotic terms only and should not depend on an electric PD.

At the ceiling the two terms can vary complementarily only. If, for instance, the concentrations of Cl^- are lower than those of Na^+, the chemical PD of the former rises faster than that of the latter. Hence at the steady state

$$RT\,\Delta\ln\,[Cl^-] \;>\; RT\,\Delta\ln\,[Na^+]$$

The opposite would be true if Cl^- were in excess of Na^+.

For an *electrogenic* Na^+ *pump*, the ceiling is determined by the electrochemical PD of Na^+. Accordingly, the chemical PD of Na^+ at the ceiling has to vary complementarily with the electric PD, and vice versa, and should therefore be affected by an electric PD change

$$RT\,\Delta\ln\,[Na^+] \;+\; F\Delta\Psi \;=\; \text{ceiling} \tag{25b}$$

However, at a fixed ceiling, the PD is a function of the chemical PD of the passive ions, here of Cl^-. At static head, the PD is even equal to the chemical PD of the passive ions as the latter must have reached complete equilibrium distribution. Hence at static head, the equation is the same as that for the electrically silent pump. But also outside the static head, though the chemical PD of Cl^- is no longer equal to the electric PD, the latter varies linearly with the former so that the chemical PD of Na^+ and Cl^- can still be treated as complementary though the ceiling is somewhat different in the two kinds of steady state. How much of a ceiling is expressed by the chemical PD of the actively transported ion, here Na^+, and how much by the electric PD, depends again on the relative concentrations of Na^+ and Cl^-, as is illustrated by the following borderline situations:

1. If there is only one salt present in the system, so that the concentrations of both the active and passive ion species are always equal in each compartment, the electrochemical potential difference of the active species, here Na^+, will be half chemical and half electrical at the ceiling.

$$RT\,\Delta\ln\,[Na^+] \;=\; -F\Delta\Psi$$

2. If the concentration of Cl-ions greatly exceeds that of Na-ions, the passive flow of Cl^- accompanying the net transport of Na-ions will make little difference to the Cl^- distribution. The ceiling of the pump will already be reached at a small change in Cl^-

distribution ratio, i.e., at a small PD change, whereas the major part of the electrochemical potential difference in the steady state will be covered by the chemical term.

$$RT\,\Delta \ln\,[Na^+] \gg -F\Delta\Psi$$

3. If the concentration of chloride is smaller than that of Na^+, even a small transfer of Cl-ions will suffice to greatly increase the distribution of these ions, and hence the electrical PD. As a result, the ceiling of the electrochemical potential of Na^+ will already be reached after little change of Na^+ distribution. The ceiling, i.e., the electrochemical potential difference of Na^+ in the steady state, will be predominantly made up by an electrical potential difference.

$$RT\,\Delta \ln\,[Na^+] \ll -F\Delta\Psi$$

The situation becomes more complicated if the Cl^- concentration is initially not equal on both sides of the membrane. If, for instance, the Cl^- concentration is initially higher on side ' than on side ", whereas that of the Na-ions is equal, a diffusion potential difference of Cl^- would already exist prior to the initiation of the electrogenic pump. As a consequence, at static head the chemical potential difference of the Na-ions may exceed the ceiling value of the electrochemical potential difference for this ion. In other words, the pump would be able to reach a seemingly higher accumulation ratio for Na-ions than would correspond to its thermodynamic ceiling. This is because the excess chemical potential difference is compensated for by a corresponding electrical potential which has the opposite orientation to those under case 1 to 3, so that the actual electrochemical potential difference for Na-ions does not exceed the ceiling set to the pump.

These considerations also apply to systems in which the second ion is also a cation, as in the K^+-H^+-exchange pump, to be discussed later.

1.3.2 Formal Treatment of Pump PD's

Transport processes can, as has been discussed in more detail elsewhere, be treated formally in two different ways: *kinetically*, i.e., in terms of the *Law of Mass Action* (LMA) or *energetically*, i.e., in terms of *Thermodynamics of Irreversible Processes* (TIP) (Heinz, 1978). The LMA expresses the transport rate as a function of the difference between the (electrochemical) activities of the permeating solute at the two sides of the separating membrane. The TIP, on the other hand, expresses the same rate as a function of the corresponding difference between the (electrochemical) potentials.

1.3.2.1 Treatment in Terms of the Law of Mass Action (LMA)

This will be applied to both an electrically silent and an electrogenic pump, using as examples some more real systems than those used for our qualitative understanding, which are of great actuality in current biological research and which are concerned with the exchange of two cation species: the electrically silent H^+-K^+-exchange pump (ATPase), which supposedly underlies the acid secretion by the gastric parietal cell (Sachs, 1977; Machen and Forte, 1979), the electrogenic (redox) H^+-pump, which operates in mitochondria (Mitchell, 1961; 1966) and various microorganisms (Rottenberg, 1975; Stoeckenius et al., 1979), and the Na^+-K^+-exchange pump (ATPase) which is partly electrically silent and partly electrogenic, and operates in muscle cells (Adrian and Slayman, 1966), nerve cells (Thomas, 1972), red blood cells (Sen and Post, 1964) and many other animal cells. For these models we make the following simplifying assumptions: (1) Only two ion species move across the membrane to an appreciable extent, so that the effect of other ions on the electrical PD can be neglected. (2) The passive movement of each of these species is exclusively rheogenic and symmetrical.

As has been discussed elsewhere, ions are likely to pass high nonelectrical energy barriers, such as occur in biological membranes, by discrete "jumps" rather than by continuous flow in free solution. Hence the influence of an electric field on these jumps may be approximately represented by the simplified transport equations [Eqs. (22a and b)], which in more general form read:

$$J_i = P_i \left(a_i' \xi^{\frac{z_i}{2}} - a_i'' \xi^{-\frac{z_i}{2}} \right)$$

z_i being the electrical valency of the ion concerned, a_i its (chemical) activity.

The general pump equation for any ion is accordingly

$$J_i = \nu_i J_r + P_i a'_i \xi^{\frac{z_i}{2}} - p_i a''_i \xi^{-\frac{z_i}{2}} \tag{26}$$

in which J_r is the pumping rate and ν_i the stoichiometric coefficient, giving number of ions of species i pumped per pumping cycle.

In the *electrically silent H^+-K^+-exchange pump,* ν_H and ν_K are equal (Fig. 6a). For reasons of simplicity, we assume that they are unity (which in reality may not be so).

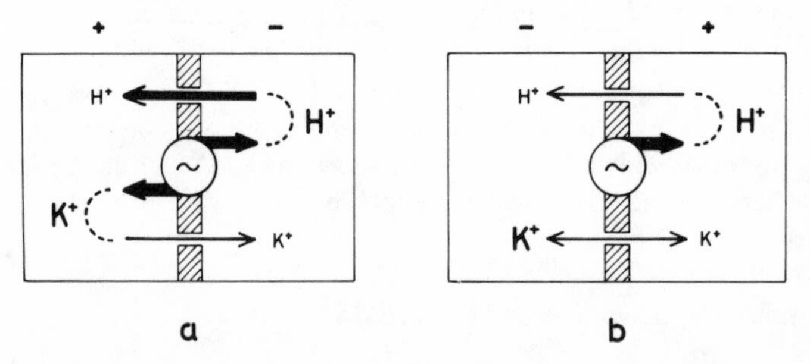

Fig. 6a, b. Ion pumps for K^+ and H^+. **a** Electrically silent pump. As indicated by the thickness of the arrows the permeability of the membrane for H-ions is greater than that for K-ions. **b** Electrogenic H^+ pump, with passive K^+ permeability. Even if the passive permeability of K^+ and H^+ were equal, an electrical potential would be generated by the H^+ pump. Note that the PD is positive on the right side, in contrast to the potential that would result from the electrically silent pump in **a**

The two equations for H^+ and K^+, respectively, as derived from Eq. (26), are

$$J_H = -\nu_H J_r + P_H [H^+]' \xi^{\frac{1}{2}} - P_H [H^+]'' \xi^{-\frac{1}{2}} \tag{27a}$$

$$J_K = \nu_K J_r + P_K [K^+]' \xi^{\frac{1}{2}} - P_K [K^+]'' \xi^{-\frac{1}{2}} \tag{27b}$$

The pumping terms must have opposite signs since the two ion species are coupled to the same process but driven in the opposite direction. As maintenance of electroneutrality requires that $J_H + J_K = 0$, the pumping terms can be eliminated so that for the transient state, ($J_H + J_K = 0$), we arrive again at the Goldman-Hodgkin-Katz equation for a membrane diffusion PD [Eq. (23)]

$$F\Delta\Psi = -RT \ln \left(\frac{P_H [H^+]'' + P_K [K^+]''}{P_H [H^+]' + P_K [K^+]'} \right)$$

For static head when all net fluxes vanish ($J_H = 0, J_K = 0$), we obtain the same equation except that the concentrations of H^+ and K^+ have reached their static head values. In contrast to the electrogenic pump to be discussed later, the electrically silent ion pump affects the electrical PD only to the extent that it influences the distribution of ion. PD maintained by an electrically silent pump is completely determined by the concentration and passive permeabilities of the transported ion species, no matter what the rate of the pump may be. Equations (27a and b) can, of course, at static head, be evaluated separately to give the electric PD as a function of pumping rate (J_r) and of the distribution if H^+ and K^+, respectively. The same electric PD will be obtained if the driving force rather than the transport is treated as constant.

We now turn to the alternative model of K^+-H^+ exchange, namely an *electrogenic proton pump* in a membrane leaky to K^+ and H^+ only. The assessment of the contribution of an electrogenic pump to the overall potential is often complicated, especially if pumping rate and pumping power are not known or difficult to measure. The difficulties can be greatly reduced by making the same simplifying assumptions that have been made previously, namely that either the pumping rate (J_r) or the (nonconjugate) driving affinity (A_{ch}) is constant. Even though we cannot rely on the validity of these assumptions, the equations derived on the basis of either one should give us some fundamental insight into the quantitative relationships between pumping rate and electrical potentials.

We start again with the assumption that the pumping rate is constant. In the general pump equation [Eq. (26)] we set for this purpose $\nu_K = 0$, as K^+ is not coupled to the transport process (Fig. 6b).

The simplified transport equations for the two ions H^+ and K^+ are accordingly

$$J_H = -J_r + P_H \xi^{\frac{1}{2}} [H^+]' - P_H \xi^{-\frac{1}{2}} [H^+]'' \tag{28a}$$

$$J_K = P_K \xi^{\frac{1}{2}} [K^+]' - P_K \xi^{-\frac{1}{2}} [K^+]'' \tag{28b}$$

As before, J_r is the pumping rate, negative by definition because it is in the outward direction, J_H and J_K are the net transport rates of H^+ and K^+, respectively. It is unimportant whether these rates refer to unit area of transporting membrane, to unit dry weight of transporting tissue, or to any other reference unit, as long as the reference unit is the same for both of them. P_K and P_H are the leakage coefficients or, more precisely, the "probabilities" of passive transfer of K^+ and H^+. ξ is the electrochemical activity coefficient $\left[= \exp\left(-z\,\frac{F\Delta\Psi}{RT}\right)\right]$.

As electroneutrality requires that

$$J_H + J_K = 0$$

the two equations can be solved for $\xi^{\frac{1}{2}}$

$$\xi^{\frac{1}{2}} = \frac{J_r}{2(P_H\,[H^+]' + P_K\,[K^+]')} + \sqrt{\frac{P_H\,[H^+]'' + P_K\,[K^+]''}{P_H\,[H^+]' + P_K\,[K^+]'} + \frac{J_r^2}{4(P_H[H^+]' + P_K\,[K^+]')^2}}$$

or, after quadrating,

$$\xi = \xi_{diff} + \frac{j^2}{2} + j\sqrt{\xi_{diff} + \frac{j^2}{4}} \tag{29}$$

in which

$$\xi_{diff} = \frac{P_H\,[H^+]'' + P_K\,[K^+]''}{P_H\,[H^+]' + P_K\,[K^+]'}$$

$$j = \frac{J_r}{P_H\,[H^+]' + P_K\,[K^+]'}$$

We see that in the transient state the total ξ is composed of diffusional terms, represented by the Goldman-Hodgkin-Katz ratio, and of pump terms, containing j. We also see that even at constant pumping rate the contribution of the pump, j, depends on the outside concentrations and passive mobilities of the two ion species. If, however, the inner space (") is small as compared to the space of the medium ('), we may neglect the changes in the outside concentrations, so that at constant J_r also j may be treated as a constant. Under these circumstances, ξ varies only with the diffusional term, to the extent that this is changed by the pumping activity during the time of observation. Clearly pump terms and diffusion terms cannot be separated from each other, i.e., neither the diffusional term nor the pump terms can influence the PD independently of the other. This is in contrast to the corresponding treatment in terms of TIP to be discussed later (1.3.2.2).

As Eq. (29) does not contain a time function, its practical usefulness is limited to a sufficiently slow or stable system. It may describe the steady state of a pump between two compartments which are so great that concentration changes can be neglected during the time of observation; for instance, in stationary flow across an epithelial layer. This equation is however less suitable for a system with rapid changes, such as in electrogenic transport into or out of a small compartment. The time function of a transient state, as will be discussed later (2.1) is rather involved, but the treatment becomes simpler again after *static head* has been reached at which the net flows of the ions concerned vanish. The corresponding equation, derived from Eq. (28a, b) simply by setting $J_H = 0, J_K = 0$, is

$$\xi^{\frac{1}{2}} = \frac{J_r}{2\,P_H\,[H^+]'} + \sqrt{\frac{[H^+]''}{[H^+]'} + \frac{J_r^2}{4\,P_H^2\,[H^+]'^2}} = \sqrt{\frac{[K^+]''}{[K^+]'}} \tag{30a}$$

or in the quadratic form

$$\xi = \frac{[K^+]''}{[K^+]'}$$

$$= \frac{[H^+]''}{[H^+]'} + \frac{J_r^2}{2\,(P_H\,[H^+]')^2} + \frac{J_r}{P_H\,[H^+]'}\sqrt{\frac{[H^+]''}{[H^+]'} + \frac{J_r^2}{4\,(P_H\,[H^+]')^2}} \tag{30b}$$

We see that the distribution of the passive K^+ reaches complete equilibrium whereas that of the pumped H^+ does not; it depends on J_r. If the pump is suddenly turned off, Eqs. (30a and b) no longer apply, since the net flows J_H and J_K are not zero any more. Hence ξ will drop according to Eq. (27), under the condition that J_r is 0, from

$$\frac{[K^+]''}{[K^+]'} \quad \text{to} \quad \frac{P_K\,[K^+]'' + P_H\,[H^+]''}{P_K\,[K^+]' + P_H\,[H^+]'}$$

Considering the low concentrations of H^+ as compared to those of K^+ in the system one may expect this PD change to be very small, provided that P_H represents only the leakage permeability of H^+. The model of constant J_r considers only this leakage pathway and does not provide for an additional pathway for H^+ equibration through the deenergized pump. By contrast, the subsequently treated model with constant driving force should at least theoretically allow the pumping pathway, after being cut off from the energy supply, to be available for facilitated diffusion of H^+ as would make the terms containing $[H^+]$ significant and the drop in PD sizeable [see Eq. (33c)].

It also follows that in the steady state the electrical potential difference, with or without an active pump, cannot exceed the value

$$RT\,\Delta \ln\,[K^+]$$

whereas in the transient state it may be considerably higher, due to the pumping.

If in such a proton pump the electroneutrality is maintained by an anion rather than by a cation, the equations are quite analogous. For example, if Cl^- were the only passively permeant ion, the corresponding terms of Eq. (29) would be

$$\xi_{diff} = \frac{P_H\,[H^+]'' + P_{Cl}\,[Cl^-]'}{P_H\,[K^+]' + P_{Cl}\,[Cl^-]''}$$

and

$$j = \frac{J_r}{P_H\,[H^+]' + P_{Cl}\,[Cl^-]''}$$

and at static head [Eq. (30)]

$$\xi = \frac{[Cl^-]'}{[Cl^-]''}$$

Treating the pumping rate J_r as a constant does not restrict the validity of the derived equations to these conditions, since J_r can be replaced by any function. We shall do so for the other assumption, namely that the *driving affinity* of the pump, rather than the pumping rate, remains constant. An approximative procedure to handle this condition is to treat the pumping process like a chemical reaction (quasi-chemical approach); consisting of two coupled processes: an osmotic, vectorial one

$$H^{+''} \rightleftharpoons H^{+'}$$

and a chemical, scalar one

$$\nu_s\, S \rightleftharpoons \nu_p\, P$$

so that

$$H^{+''} + \nu_s\, S \rightleftharpoons H^{+'} + \nu_p\, P$$

S and P being substrate and product of the driving chemical reaction, ν_s and ν_p the corresponding stoichiometric coefficients. For simplicity reasons we assume only one species of S and P, respectively is involved. If more species were involved in the driving reaction, the equation could easily be expanded accordingly.

The rate of the overall coupled process, J_r, would be, as for a chemical reaction, equal to

$$J_r = k_1 \tilde{a}''_h \cdot s^{\nu_s} - k_{-1} \tilde{a}'_h \cdot p^{\nu_p}$$

$\tilde{a}_h$ representing the electrochemical activity of H^+, s and p the concentrations of substrate S and product P, respectively. Factorizing $k_{-1} \cdot p^{\nu_p}$, we obtain

$$J_r = k_{-1}\, p^{\nu_p} \left(\frac{k_1}{k_{-1}} \cdot \frac{s^{\nu_s}}{p^{\nu_p}}\, \tilde{a}''_h - \tilde{a}'_h \right)$$

Replacing $k_{-1}\, \mathrm{p}^{\nu_\mathrm{p}}$ by k_r, the overall rate coefficient, and $\frac{k_1}{k_{-1}} \cdot \frac{\mathrm{s}^{\nu_\mathrm{s}}}{\mathrm{p}^{\nu_\mathrm{p}}}$ by Γ, the "chemical reactivity coefficient" (Heinz, 1978), we obtain

$$J_\mathrm{r} \;=\; k_\mathrm{r}\; \Gamma \tilde{a}''_\mathrm{h} \;-\; \tilde{a}'_\mathrm{h} \tag{31}$$

Since $\Gamma = \exp\ (\frac{A_\mathrm{ch}}{RT})$, it will be constant because A_ch, the affinity of the driving reaction, is assumed to be constant in the present approach. Inserting Eq. (31) into Eq. (23a), we obtain the basic equation

$$J_\mathrm{H} \;=\; k_\mathrm{r}\; ([\mathrm{H}^+]'\, \xi^{\frac{1}{2}} \;-\; \Gamma\, [\mathrm{H}^+]''\, \xi^{-\frac{1}{2}}) \;+\; P_\mathrm{H}\; ([\mathrm{H}^+]'\, \xi^{\frac{1}{2}} \;-\; [\mathrm{H}^+]''\, \xi^{-\frac{1}{2}}) \tag{32}$$

whereas that for K^+ is the same as Eq. (28b). It should be pointed out that k_r is itself a function of Γ; it usually becomes smaller with increasing Γ. If we in the following treat k_r as constant, we oversimplify the system, but the dependence of k_r on Γ is presumably small enough to be neglected for the present purpose.

Again, we look at the PD at two different states of the system: (1) the transient or stationary state, assuming continuous net movement of both K^+ and H^+ under electroneutrality condition, and (2) the static head, assuming that net movement of each H^+ and K^+ has vanished.

1. In the transient state or at stationary flow:

$$F\Delta\Psi \;=\; -RT \ln \frac{(k_\mathrm{r}\, \Gamma \;+\; P_\mathrm{H})\, [\mathrm{H}^+]'' \;+\; P_\mathrm{K}\, [\mathrm{K}^+]''}{(k_\mathrm{r} \;+\; P_\mathrm{H})\, [\mathrm{H}^+]' \;+\; P_\mathrm{K}\, [\mathrm{K}^+]'} \tag{33a}$$

2. At static head:

$$F\Delta\Psi \;=\; -RT \ln \frac{(k_\mathrm{r}\, \Gamma \;+\; P_\mathrm{H})\, [\mathrm{H}^+]''}{(k_\mathrm{r} \;+\; P_\mathrm{H})\, [\mathrm{H}^+]'} \;=\; -RT \ln \frac{[\mathrm{K}^+]''}{[\mathrm{K}^+]'} \tag{33b}$$

The residual PD after sudden interruption of energy supply ($\Gamma = 1$) should be

$$F\Delta\Psi \;=\; -RT \ln \frac{(P_\mathrm{H} \;+\; k_\mathrm{r})\, [\mathrm{H}^+]'' \;+\; P_\mathrm{K}\, [\mathrm{K}^+]''}{(P_\mathrm{H} \;+\; k_\mathrm{r})\, [\mathrm{H}^+]' \;+\; P_\mathrm{K}\, [\mathrm{K}^+]'} \tag{33c}$$

which might indicate a considerable drop or even inversion of the PD in either state (Heinz, 1981). As mentioned before, this drop does not follow from the corresponding equations derived for the model assuming constant pumping rate. If it occurs it is probably of short duration, as, depending on the mechanics of energetic coupling in the system, k_r is likely to decrease or vanish sooner or later, so that Eq. (33c) will be transformed into the Goldman-Hodgkin-Katz equation, and the subsequent diffusional decay of the electric PD will be much slower.

To derive the analogous equations for an *electrogenic exchange pump* (active antiport), e.g., for the *Na-K-pump* in red blood cells (Fig. 7), we start with individual flow

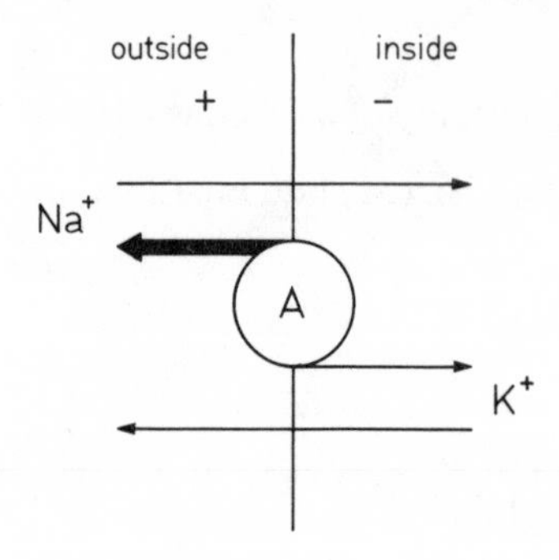

Fig. 7. Electrogenic Na-K-exchange pump. The thickness of the arrows indicates that the number of Na^+-ions pumped with each cycle is greater than the number of K-ions

equations, as above, except that now both equations contain the pump term, but with different stoichiometric coefficients and signs:

$$J_{Na} = -\nu_{Na} J_r + P_{Na} [Na^+]' \xi^{\frac{1}{2}} - P_{Na} [Na^+]'' \xi^{-\frac{1}{2}} \tag{34a}$$

$$J_K = \nu_K J_r + P_K [K^+]' \xi^{\frac{1}{2}} - P_K [K^+]'' \xi^{-\frac{1}{2}} \tag{34b}$$

As Na^+ is pumped from " to ', $\nu_{Na} J_r$ is taken to be negative. We first assume that the pumping rate J_r is constant. As the net flows of the two ions must be opposite and equal, we can proceed as before by solving the equations for ξ and obtain an equation analogous to that for active uniport [Eq. (29)]:

$$\xi = \xi_{diff} + \frac{j^2}{2} + j \sqrt{\xi_{diff} + \frac{j^2}{4}} \tag{35a}$$

but this time

$$\xi_{diff} = \frac{P_{Na} [Na^+]'' + P_K [K^+]''}{P_{Na} [Na^+]' + P_K [K^+]'}$$

$$j = \frac{(\nu_{Na} - \nu_K) J_r}{P_{Na} [Na^+]' + P_K [K^+]'}$$

For similar reasons as for Eq. (29), we may consider j, the term of the electrogenic pump, constant during the time of observation, whereas the diffusional term, ξ_{diff}, gradually changes during the pump activity. Also this equation does not tell us the time after which a steady state is reached, as it has no time function.

By setting the net flows of both Na^+ and K^+ [Eqs. (34a and b)] equal to zero, we arrive at the well-known steady state equation by Mullins and Noda (1963):

$$\xi_{ss} = \frac{\nu_{Na} P_K [K^+]'' + \nu_K P_{Na} [Na^+]''}{\nu_{Na} P_K [K^+]' + \nu_K P_{Na} [Na^+]'} = \frac{r P_K [K^+]'' + P_{Na} [Na^+]''}{r P_K [K^+]' + P_{Na} [Na^+]'} \quad (35b)$$

$r = \dfrac{\nu_{Na}}{\nu_K}$, the stoichiometry of the pump with respect to Na^+ and K^+ exchange per cycle.

If the pump is turned off at this stage, ξ will drop to the corresponding value given by the Goldman-Hodgkin equation for the same concentrations:

$$\xi_{diff} = \frac{P_K [K^+]'' + P_{Na} [Na^+]''}{P_K [K^+]' + P_{Na} [Na^+]'}$$

This may not be a great drop if the Na^+ terms are much smaller than the K^+ terms, provided that leakage is by free diffusion only, as appears to be implied at constant J_r. As has already been mentioned for the previous system, this does not necessarily apply for the subsequent treatment at constant driving force if the pumping mechanism allows leakage by facilitated diffusion [Eq. (39)]. It follows from Eq. (35) that at static head the electrical potential difference cannot exceed

$$-RT \Delta \ln [K^+]$$

but in the transient state or at steady flow it might very much exceed this value, depending on the pumping rate (A'drian and Slayman, 1966; Pietrzyk et al., 1978).

The Na-K-exchange pump can also be treated under the alternative assumption that the driving affinity (A_{ch}) rather than the rate of the pump remains constant. In analogy to the previous treated H^+ pump we can write the pumping process as a chemical reaction:

$$\nu_{Na} Na^{+''} + \nu_K K^{+'} + \nu_s S \rightleftharpoons \nu_{Na} Na^{+'} + \nu_K K^{+''} + \nu_p P$$

S and P representing substrate(s) and product(s) of the driving reaction.

The rate of the overall pumping process would then be

$$J_r = k_1 (\widetilde{a}_{Na}{}^{\nu_{Na}})'' \cdot (\widetilde{a}_K{}^{\nu_K})' \cdot s^{\nu_s} - k_{-1} (\widetilde{a}_{Na}{}^{\nu_{Na}})' \cdot (\widetilde{a}_K{}^{\nu_K})'' \cdot p^{\nu_p}$$

Again, we have for simplicity reasons assumed that only one species of each substrate and product, respectively, is involved. By factorizing $k_{-1} \cdot p^{\nu_p}$ we obtain

$$J_r = k_{-1} p^{\nu_p} \left[\frac{k_1}{k_{-1}} \cdot \frac{s^{\nu_s}}{p^{\nu_p}} \cdot (\widetilde{a}_{Na}{}^{\nu_{Na}})'' \cdot (\widetilde{a}_K{}^{\nu_K})' - (\widetilde{a}_{Na}{}^{\nu_{Na}})' \cdot (\widetilde{a}_K{}^{\nu_K})'' \right] \quad (36)$$

and after replacing $k_{-1}\,p^{\nu_p}$ by k_r, $\frac{k_1}{k_{-1}}\,\frac{s^{\nu_s}}{p^{\nu_p}}$ by Γ, and the electrochemical activities of the cations on the left side (') by $[Na^+]'\,\xi^{\frac{1}{2}}$ and $[K^+]'\,\xi^{\frac{1}{2}}$, respectively, and on the right side (") by $[Na^+]''\,\xi^{-\frac{1}{2}}$ and $[K^+]''\,\xi^{-\frac{1}{2}}$, respectively, we can, in analogy to the treatment of the H^+ pump, insert the above expression (36) for J_r in Eqs. (34a and b). If we assume that $\nu_{Na} = 3$ and $\nu_K = 2$ (Sen and Post, 1964), we obtain for the individual ion flows

$$J_{Na} = 3\,k_r\,\left\{([Na^+]^3)''\cdot([K^+]^2)'\,\xi^{-\frac{1}{2}}\,\Gamma - ([Na^+]^3)'([K^+]^2)''\,\xi^{\frac{1}{2}}\right\} + P_{Na}\,([Na^+]'\,\xi^{\frac{1}{2}} - [Na^+]''\,\xi^{-\frac{1}{2}}) \tag{37a}$$

$$J_K = 2\,k_r\,\left\{([Na^+]^3)''\cdot([K^+]^2)'\,\xi^{-\frac{1}{2}}\,\Gamma - ([Na^+]^3)'([K^+]^2)''\,\xi^{\frac{1}{2}}\right\} + P_K\,([K^+]'\,\xi^{\frac{1}{2}} - [K^+]''\,\xi^{-\frac{1}{2}}) \tag{37b}$$

Considering that the maintenance of electroneutrality requires that

$$J_{Na} + J_K = 0$$

and assuming that the rheogenic Cl^- movement be negligible, we can combine the equations to solve for ξ:

$$\xi = \frac{P_{Na}\,[Na^+]'' + P_K\,[K^+]'' + k_r\,([Na^+]'')^3\cdot([K^+]')^2\,\Gamma}{P_{Na}\,[Na^+]' + P_K\,[K^+]' + k_r\,([Na^+]')^3\cdot([K^+]'')^2} \tag{38}$$

At static head J_{Na} and J_K each vanishes so that from Eqs. (37a) and (37b), respectively, we get

$$\xi = \frac{P_{Na}\,[Na^+]'' + 3\,k_r\,\Gamma\,([Na^+]'')^3\cdot([K^+]')^2}{P_{Na}\,[Na^+]' + 3\,k_r\,([Na^+]')^3\cdot([K^+]'')^2}$$

$$= \frac{P_K\,[K^+]'' - 2\,k_r\,\Gamma\,([Na^+]'')^3\cdot([K^+]')^2}{P_K\,[K^+]' - 2\,k_r\,([Na^+]')^3\cdot([K^+]'')^2}$$

Combining these equations we obtain

$$\xi = \frac{2\,P_{Na}\,[Na^+]'' + 3\,P_K\,[K^+]''}{2\,P_{Na}\,[Na^+]' + 3\,P_K\,[K^+]'}$$

which is identical with Eq. (35) for the special case that $r = \frac{3}{2}$, obtained in the previous treatment of this system under the assumption of constant pumping rate.

It should be kept in mind that the above equations are based on the simplifying assumption that all leakages involve free ions only. This assumption is reasonable as long as the pump acts exclusively as a current generator (J_r = constant). If, however, the pump is stopped, e.g., by blocking or uncoupling the supply of energy, the pumping system might allow additional leakage for the pumped ion, parallel to that of the pathways considered in the Goldman-Hodgkin-Katz ratios. In other words, the system of active ion transport might be converted into one of facilitated diffusion, as has been shown to occur with some systems in microorganisms which have lost the active transport component by mutation (Okada and Halvorson, 1964) or by specific inhibition (Kashket and Wilson, 1972; Komor et al., 1974). For the present Na^+-K^+-exchange pump, this could be expressed formally by Eq. (38a) under the condition that the nonconjugate energy source is zero ($\Gamma = 1$):

$$\xi = \frac{P_{Na}\,[Na^+]'' + P_K\,[K^+]'' + k_r\,[Na^+]''^3\,[K^+]'^2}{P_{Na}\,[Na^+]' + P_K\,[K^+]' + k_r\,[Na^+]'^3\,[K^+]''^2} \tag{39}$$

Under these conditions, turning off the pump ($\Gamma \rightarrow 1$) might cause a rapid and significant drop in ξ from the static head value to that indicated by Eq. (39). It is, however, still uncertain whether this backflow of ions through an inactive pump system has a significant effect on the PD.

In all previously treated cases, the contribution of the pump to the overall PD, i.e., the difference between the latter and the residual PD obtained immediately after turning off the pump, becomes smaller as the system approaches the steady state. This is so even if either the pumping rate (J_r) or the pumping affinity (A_{ch}) remains the same while the pump is active. This follows from the fact that the contribution of the pump, f(j) is added to ξ_{diff} and not to the PD, so that

$$F\Delta\Psi_{pump} = -RT \ln \frac{\xi_{diff} + f(j)}{\xi_{diff}} \tag{40}$$

1.3.2.2 Treatment in Terms of Thermodynamics of Irreversible Processes (TIP)

The electrical PD as a function of an electrogenic pump and of the concentrations and mobilities of permeant ions can also be expressed in terms of TIP. The resulting equations are much simpler and more lucid than the corresponding kinetic ones, but are valid only to the extent that the phenomenological coefficients (L_i) can be treated as constant parameters with sufficient approximation. This is generally true if the system is close enough to equilibrium, but also for distinct domains far away from equilibrium (Rottenberg, 1973; Heinz, 1978). In any case the constancy of the L values has to be tested experimentally, before applying the notation of irreversible thermodynamics. It should be emphasized, however, that the TIP procedure is applicable whenever the widely used linear electrical circuitry is applicable, since both approaches depend on the same assumptions and differ otherwise only in their units.

We shall begin with the same active transport systems as treated previously, namely those which lead to an exchange between K^+ and H^+. As in the previous chapter this could be either an electrically silent H^+-K^+-exchange pump, or an electrogenic H^+-pump under the condition that K^+ is the only other permeant ion species present. The general phenomenological equations applied to the H^+-K^+-exchange pump, when considered to be coupled to an exergonic chemical reaction can be written as follows:

$$J_H = L_{HH} X_H + L_{HK} X_K + L_{Hc} A_{ch} \tag{41a}$$

$$J_K = L_{HK} X_H + L_{KK} X_K + L_{Kc} A_{ch} \tag{41b}$$

$$J_{ch} = L_{Hc} X_H + L_{Kc} X_K + L_{cc} A_{ch} \tag{41c}$$

J_H and J_K are the net flows of H^+ and K^+, respectively, J_{ch} the rate of the (driving) chemical reaction. X_H and X_K are the (negative) electrochemical potential differences of H^+ and K^+, respectively. A_{ch} is the affinity of the driving reaction. L_{HH}, L_{KK}, L_{cc}, L_{HK}, L_{HC}, and l_{Kc} are the phenomenological coefficients. A more thorough treatment of active transport in these terms has been given by Essig and Caplan (1968). For a stoichiometrically coupled process we prefer the quasi-chemical notation which treats the two coupled processes, the osmotic and the chemical one, as a single overall reaction of fixed stoichiometry (Heinz, 1978). As in chemical equations only the initial and final states of these reactions, corresponding to reactants and products, are given. To apply this to the K^+-H^+-exchange pump, the coupled processes are (1) the translocation of ν_H H-ions from compartment to compartment and of ν_K K-ions in the opposite direction, and (2) a (driving) chemical reaction which for the sake of simplicity is assumed to transform ν_S molecules of the substrate S to form ν_p molecules of the product P. The overall coupled reaction is therefore

$$\nu_H H^{+''} + \nu_K K^{+'} + \nu_S \dot{S} \rightarrow \nu_H H^{+'} + \nu_K K^{+''} + \nu_P P$$

The overall affinity, A_r, of this reaction at each stage is composed of the differences in electrochemical potential of each participating process or reaction

$$A_r = \nu_H \Delta\tilde{\mu}_{H^+} - \nu_K \Delta\tilde{\mu}_{K^+} - (\nu_p \mu_p - \nu_S \mu_S)$$

Introducing conventional symbols X_{H^+} and X_{K^+} for $-\Delta\tilde{\mu}_{H^+}$ and $-\Delta\tilde{\mu}_{K^+}$, respectively, and A_{ch}, the affinity of the (driving) chemical reaction, for

$$(\nu_s \mu_s - \nu_p \mu_p)$$

we obtain

$$A_r = -\nu_H X_H + \nu_K X_K + A_{ch}$$

The flow of the overall coupled process is

$$J_r = L_r A_r = L_r(-\nu_H X_H + \nu_K X_K + A_{ch}) \tag{42}$$

The total flow of each species includes (uncoupled) leakage flows, which for simplicity reasons are treated as linear functions of the corresponding electrochemical potential differences, or affinities, respectively:

$$J_H = -\nu_H J_r + L_H^u X_H \tag{43a}$$

$$J_K = \nu_K J_r + L_K^u X_K \tag{43b}$$

$$J_{ch} = J_r + L_{ch}^u A_{ch} \tag{43c}$$

L_H^u, L_K^u, L_{ch}^u being the proportionality constants of the (uncoupled) leakage flows of H^+, K^+, and of the chemical reaction, respectively. J_r can be replaced according to Eq. 24.

The complete phenomenological equations are

$$J_H = (\nu_H^2 L_r + L_H^u) X_H - \nu_H \nu_K L_r X_K - \nu_H L_r A_{ch} \tag{44a}$$

$$J_K = -\nu_H \nu_K L_r X_H + (\nu_K^2 L_r + L_K^u) X_K + \nu_K L_r A_{ch} \tag{44b}$$

$$J_{ch} = -\nu_H L_r X_H + \nu_K L_r X_K + (L_r + L_{ch}^u) A_{ch} \tag{44c}$$

It can be seen that these equations become identical with Eqs. (41a, b, c) if one sets:

$$\nu_H^2 L_r + L_H^u = L_{HH}$$

$$\nu_K^2 L_r + L_K^u = L_{KK}$$

$$L_r + L_{ch}^u = L_{cc}$$

$$\nu_H \nu_K L_r = L_{HK}$$

$$\nu_H L_r = L_{Hc}$$

$$\nu_K L_r = L_{Kc}$$

Each of the electrochemical potential differences can be split into a chemical and an electric term:

$$X_H = -\Delta\tilde{\mu}_H = -RT\,\Delta \ln [H^+] - F\Delta\Psi \tag{45a}$$

$$X_K = -\Delta\tilde{\mu}_K = -RT\,\Delta \ln [K^+] - F\Delta\Psi \tag{45b}$$

In order to apply the above equations to an *electrically silent H^+-K^+ pump*, we set $\nu_H = \nu_K$. If for simplicity reasons we consider them as unity (which is not necessarily true), and as both cations are monovalent, we derive from Eqs. (44a and b), in combination with Eqs. (45a and b), the electric PD under transient state or stationary flow conditions. As for electroneutrality reasons, in the absence of appreciable other ion flows.

$$J_H + J_K = 0:$$

From Eqs. 43a and b we obtain

$$X_K - X_H = -\frac{L_H^u + L_K^u}{L_K^u L_H^u}(J_r - J_{net}) \tag{46a}$$

or, as the electric terms of X_K and X_H cancel,

$$RT\,\Delta\ln[K^+] - RT\,\Delta\ln[H^+] = \frac{L_H^u + L_K^u}{L_H^u L_K^u}(J_r - J_{net}) \tag{46b}$$

J_{net} being the (transient) net exchange rate ($= -J_H$). $X_K - X_H$ can be considered a (transient) *ceiling*, as long as J_r and J_{net} are fixed. The electric PD under these conditions is

$$F\Delta\Psi = -\frac{L_H^u RT\,\Delta\ln[H^+] + L_K^u RT\,\Delta\ln[K^+]}{L_H^u + L_K^u} \tag{47a}$$

At *static head*, when both J_K and J_H vanish, we obtain the *maximum ceiling*

$$(X_K - X_H)_{max} = RT\,\Delta\ln[H^+]_s - RT\,\Delta\ln[K^+]_s = \frac{L_H^u + L_K^u}{L_H^u L_K^u}J_r \tag{48}$$

The electric PD under these conditions is

$$F\Delta\Psi_s = -\frac{L_H^u RT\,\Delta\ln[H^+]_s + L_K^u\,\Delta\ln[K^+]_s}{L_H^u + L_K^u} \tag{47b}$$

This equation is formally the same as that for the transient PD, except that H^+ and K^+ now have their static head concentrations. It follows that $F\Delta\Psi$ does not directly depend on J_r, but is fully determined by the distribution ratios of H^+ and K^+ at either state.

These equations confirm at least qualitatively the conclusions arrived at by the procedure according to LMA.

To apply the analogous procedure to the *electrogenic proton pump*, we have to set $\nu_K = 0$. The phenomenological equations become:

$$J_H = (\nu_H^2 L_r + L_H^u)X_H - \nu_H L_r A_{ch} \tag{49a}$$

$$J_K = L_K^u X_K \tag{49b}$$

The further procedure can again be greatly simplified by assuming that either J_r, the pumping rate, or A_{ch}, the driving affinity, is constant during the experiment. Starting with constant J_r we obtain

$$J_H = -\nu_H J_r + L_H^u (-RT \Delta \ln [H^+] - F\Delta\Psi) \tag{50a}$$

and

$$J_K = L_K^u (-RT \Delta \ln [K^+] - F\Delta\Psi) \tag{50b}$$

Setting again $J_H + J_K = 0$, we obtain for the transient state or the stationary flow

$$-\nu_H J_r - L_H^u (RT \Delta \ln [H^+] + F\Delta\Psi) = -L_K^u (RT \Delta \ln [K^+] + F\Delta\Psi)$$

For stationary flow it follows that at constant J_r

$$X_H = \frac{\nu_H J_r - J_{net}}{L_H^u}$$

which represents a "ceiling" provided that the net output ($J_{net} = -J_H$) is fixed. The electric PD under these conditions is

$$F\Delta\Psi = -\frac{\nu_H J_r + L_H^u RT \Delta \ln [H^+] + L_K^u RT \Delta \ln [K^+]}{L_H^u + L_K^u} \tag{51}$$

It is seen that the electric PD is no longer determined by the distribution of H^+ and K^+ alone, but gets an explicit contribution from the pump. At static head, each J_H and J_K becomes zero, so that

$$F\Delta\Psi_s = -RT \Delta \ln [K^+] = -\nu_H \frac{J_r}{L_H^u} - RT \Delta \ln [H^+] \tag{52}$$

and the electrochemical potential difference of H^+ (ceiling) at static head is

$$(X_H)_{max} = -\Delta\tilde{\mu}_H = -(F\Delta\Psi_s + RT \Delta \ln [H^+]) = \frac{\nu_H J_r}{L_H^u} \tag{53}$$

combining Eqs. (53) and (52) we get for the maximum ceiling

$$(X_H)_{max} = -(RT \Delta \ln [H^+] + RT \Delta \ln [K^+]) = \frac{\nu_H J_r}{L_H^u} \tag{54}$$

showing that at static head the difference between the chemical PD's equals the ceiling.

As had been predicted qualitatively before, with excess of $[K^+]$ the comparatively small amount of H^+ transported toward the ceiling will hardly change the K^+ distribution and hardly the electric PD. As a consequence, the chemical PD of H^+ covers the major fraction of the "electron motive force" (EMF). If on the other hand $[K^+]$ is small, the PD rises rapidly with little transportation of H^+: in this case the electrical PD will cover the major fraction of the EMF.

If at any time during the transport process the H^+ pump is suddenly turned off, the overall PD and hence the protonmotive force drops according to the above equations by the term:

$$\frac{\nu_H J_r}{L_H^u + L_K^u} \tag{55}$$

In other words, the contribution of the pump to the overall PD seems to be the same in the transient and in the steady state. In reality, however, this contribution is presumably decreasing accordingly as the system approaches the steady state, since the sum L_H^u + L_K^u can hardly remain stable but is likely to increase under these conditions.

If, instead of J_r, the affinity of the driving chemical reaction, A_{ch}, is assumed to remain constant, then the corresponding equations, which can be derived in analogy to the preceding ones, are in the *transient state* and *stationary flow* for the ceiling

$$X_H = \frac{\nu_H L_r A_{ch} - J_{net}}{\nu_H^2 L_r + L_H^u} \tag{56a}$$

and for the PD

$$F\Delta\Psi = \frac{-\nu_H L_r A_{ch} + (\nu_H^2 L_r + L_H^u) RT \Delta \ln [H^+] + L_K^u RT \Delta \ln [K^+]}{\nu_H L_r + L_H^u + L_K^u} \tag{56b}$$

At static head ($J_H = 0, J_K = 0$) the corresponding equations reduce to

$$(X_H)_{max} = RT \Delta \ln [K^+] - RT \Delta \ln [H^+] = \frac{\nu_H L_r}{\nu_H^2 L_r + L_H^u} A_{ch} \tag{57a}$$

for the maximum ceiling, and for the PD to

$$F\Delta\Psi_s = -\left(\frac{\nu_H L_r}{\nu_H^2 L_r + L_H^u} A_{ch} + RT \Delta \ln [H^+]\right) \tag{57b}$$

Also here, the contribution of the electrogenic pump to the overall PD is seemingly the same under all conditions, namely $\frac{\nu_H L_r}{\nu_H^2 L_r + L_H^u + L_K^u} A_{ch}$. But for reasons similar to those given for the condition of constant J_r, this contribution is likely to decrease as the system approaches the steady state, as the phenomenological coefficients cannot be expected to be constant over such a wide range.

To apply analogous procedure to the electrogenic Na^+-K^+-exchange pump, we start again with the equation of the coupled overall reaction,

$$\nu_{Na}\, Na^{+''} + \nu_K\, K^{+'} + \nu_S\, S \rightleftharpoons \nu_{Na}\, Na^{+'} + \nu_K\, K^{+''} + \nu_P\, P$$

S and P being substrate and product, respectively, of the driving chemical reaction. The overall rate of this reaction is

$$J_r = L_r A_r$$

$$A_r = -\nu_{Na} X_{Na} + \nu_K X_K + A_{ch}$$

A_r is again the affinity of the overall process, and A_{ch} that of the chemical reaction proper:

$$\nu_s\, S \rightleftharpoons \nu_p\, P$$

$$A_{ch} = -(\nu_P\, \Delta\mu_P - \nu_S\, \Delta\mu_S)$$

The flows of the individual ion species are:

$$J_{Na} = -\nu_{Na} J_r + L_{Na}^u X_{Na} \tag{58a}$$

$$J_K = +\nu_K J_r + L_K^u X_K \tag{58b}$$

L_{Na}^u and L_K^u being the leakage coefficients of Na^+ and K^+, respectively. To simplify the procedure we assume that either the overall reaction rate, J_r, or the driving affinity, A_{ch}, remains constant. Let us begin with the first alternative. If we split X_{Na} and X_K each into chemical and electrical terms, the equations of the individual ion fluxes become:

$$J_{Na} = -\nu_{Na} J_r + L_{Na}^u \left(-RT \ln \frac{[Na]''}{[Na]'} - F\Delta\Psi\right) \tag{59a}$$

$$J_K = \nu_K J_r + L_K^u \left(-RT \ln \frac{[K]''}{[K]'} - F\Delta\Psi\right) \tag{59b}$$

Electroneutrality requires that

$$J_{Na} + J_K = 0$$

as the mobility of other ions is assumed to be negligible in our model. We can now solve the equations for the electrical potential difference and obtain

$$F\Delta\Psi = -\frac{\nu_{Na}-\nu_K}{L^u_{Na}+L^u_K}\,J_r - \frac{RT}{L^u_{Na}+L^u_K}\,(L^u_{Na}\,\Delta\ln[Na^+] + L^u_K\,\Delta\ln[K^+]) \tag{60a}$$

Also here, $F\Delta\Psi$ is composed of two terms, one relating to the pumping rate and the other to the passive membrane diffusion potential. During the pumping activity concentrations of Na^+ and K^+ will slowly change toward their steady-state values. The static head potential difference, as each J_{Na} and J_K equals zero, becomes

$$F\Delta\Psi_s = -\frac{RT}{\nu_{Na}\,L^u_K + \nu_K\,L^u_{Na}}\,(\nu_{Na}\,L^u_K\,\Delta\ln[K^+] + \nu_K\,L^u_K\,\Delta\ln[Na^+]) \tag{60b}$$

It is seen that at static head the electrical PD does not explicitly depend on the pumping rate, in analogy to the equation of Mullins and Noda [Eq. (35)], whereas the distribution ratios of the ion species' do:

$$RT\,\Delta\ln[Na^+] = -\nu_{Na}\,\frac{J_r}{L^u_{Na}} - F\Delta\Psi_s$$

$$RT\,\Delta\ln[K^+] = \frac{\nu_K\,J_r}{L^u_K} - F\Delta\Psi_s$$

$$RT\,(\Delta\ln[Na^+] - \Delta\ln[K^+]) = -\left(\frac{\nu_{Na}}{L^u_{Na}} + \frac{\nu_K}{L^u_K}\right)J_r = (X_K - X_{Na})_{max} \tag{61}$$

It follows that at static head the electrochemical potential difference for each ion species is determined by rate and stoichiometry of the pump and by the leakage coefficients. To the extent that the latter are constant, variations of the chemical potential difference of both Na^+ and K^+ must be complementary with those of the corresponding electrical PD. Equation (61) also shows that the maximum ceiling $(X_K - X_{Na})$ is fixed under these conditions.

In order to obtain the corresponding equations for constant driving affinity (A_{ch}), we replace J_r according to Eq. (42). After rearrangement of terms the flow equations of the individual ion flows become:

$$J_{Na} = (\nu^2_{Na}\,L_r + L^u_{Na})\,X_{Na} - \nu_{Na}\,\nu_K\,L_r\,X_K - \nu_{Na}\,L_r\,A_{ch} \tag{62a}$$

$$J_K = -\nu_{Na}\,\nu_K\,L_r\,X_{Na} + (\nu^2_K\,L_r + L^u_K)\,X_K + \nu_K\,L_r\,A_{ch} \tag{62b}$$

Setting $J_{Na} + J_K = 0$ and splitting X_{Na} and X_K into the chemical and electrical term, we can solve for the electrical PD:

$$F\Delta\Psi = -\frac{(\nu_{Na} - \nu_K)\,\nu_{Na}\,L_r + L^u_{Na}}{\Sigma}\,RT\,\Delta\ln[Na^+]$$

$$+\frac{(\nu_{Na} - \nu_K)\,\nu_K\,L_r\,L^u_K}{\Sigma}\,RT\,\Delta\ln[K^+] - \frac{(\nu_{Na} - \nu_K)\,L_r}{\Sigma}\,A_{ch} \tag{63}$$

$$\Sigma = -(\nu_{Na} - \nu_K)^2\,L_r + L^u_{Na} + L^u_K$$

At static head both J_{Na} and J_K become equal to zero. So we obtain

$$F\Delta\Psi_s = -\frac{\nu_K\,L^u_{Na}\,RT\,\Delta\ln[Na^+] + \nu_{Na}\,L^u_K\,RT\,\Delta\ln[K^+]}{\nu_K\,L^u_{Na} + \nu_{Na}\,L^u_K} \tag{64}$$

which is identical with the equation obtained under the assumption that the pumping rate J_r is constant [Eq. (60b)]. It shows likewise that $\Delta\Psi$ does not explicitly depend on the driving affinity, A_{ch}, whereas the electrochemical potential differences of Na^+ and K^+ respectively, do so:

$$X_{Na} = \frac{\nu_{Na}\,L_r\,L^u_K}{L^u_{Na}\,L^u_K + L_r\,(\nu^2_{Na}\,L_K + \nu^2_K\,L^u_{Na})}\,A_{ch} \tag{65a}$$

and

$$X_K = -\frac{\nu_K L_r L^u_{Na}}{L^u_{Na}\,L^u_K + L_r\,(\nu^2_{Na}\,L^u_K + \nu^2_K\,L^u_{Na})}\,A_{ch} \tag{65b}$$

From these we obtain for the maximum ceiling $(X_K - X_{Na})_{max}$ or

$$(X_K - X_{Na})_{max} = -\frac{L_r\,(\nu_{Na}\,L^u_K + \nu_K\,L^u_{Na})}{L^u_{Na}\,L^u_K + L_r\,(\nu^2_{Na}\,L^u_K + \nu^2_K\,L^u_{Na})}\,A_{ch} \tag{65c}$$

It is seen that at constant driving affinity (A_{ch}), provided that all phenomenological and stoichiometric coefficients are also constant, the maximum electrochemical potential difference of Na^+ and K^+ and hence also the maximum ceiling is fixed, as was the case with constant pumping rate.

The equations so far derived for the electric PD in the presence of ion pumps, whether in terms of LMA or TIP, do not contain a function of time. They are useful mainly for time-independent states, such as stationary flow and static head, but they tell us little about the transition of the PD during changing conditions, for instance, after a pump is turned "on" or "off". For special biological questions, the time curve of such transitions is important, since after drastic changes in pump activity the electric PD is likely to respond much faster than the distribution of the ions involved. A rapid electric response would make pumping energy available (or unavailable) for useful functions more swiftly, and well in advance of significant changes in chemical PD. This question is of special interest for the chemiosmotic energy transduction with electrogenic proton pumps, e.g., in mitochondria, chloroplasts, bacteriorhodopsin systems, etc. The corresponding time curves are complicated by transient deviations from electroneutrality. Preliminary equations of this kind, based on simplifying assumptions, have been carried out for mitochondria by Mitchell (1968), and extended by Heinz (1981). They will be dealt with more extensively and formally later in this booklet (see 2.1). At the present juncture, a qualitative consideration may suffice provide a basic understanding.

Whereas the static-head PD does not depend on the permeability of the passive ion, the transient-state PD does, e.g. after the electrogenic pump is turned on (or off). We illustrate this by two borderline cases:

1. The passive ion movement is much faster than the pumping rate. In this case the distribution of the passive ion will be almost instantaneously in equilibrium with the PD at all times, and the overall rate is determined by the pumping rate. The electrical PD will rise slowly and concomitantly with the chemical potential difference of the active ion toward its static-head value, which depends on the relative concentrations only.

2. The passive ion movement is much slower than the Na^+ pumping rate. In this case the pump almost acts as if no permeant anions were present. Since only very few ions have to be transported in order to reach the ceiling, i.e., to charge the static capacity of the membrane, the electrical PD across the membrane will jump at once to a high value, before the distribution of the passive ion can significantly change. But as the pump slowly raises the chemical PD of the active ion, the electric PD has to decline concomitantly until it reaches its static-head value, which depends on the relative concentrations only. This will be derived more quantitatively in a later section (2.1).

A more elaborate treatment of electrogenic pumps has been presented at several places (Rottenberg 1979, Westerhoff and van Dam 1979, and others).

A comparison between the two approaches discussed, namely LMA and TIP, shows the advantages of the latter over the former. The TIP equations, besides being simpler and more lucid than the corresponding LMA equations, allow a clear separation between pump terms and diffusion terms, as we have seen. In addition, they provide us with a basis for rather simple experimental tests of whether a given pump system is electrogenic or electrically silent. Such tests will be discussed in the next paragraph. A further valuable advantage will become apparent when we later develop a time function of the electrical and chemical PD, for instance, after an electrogenic pump has been turned on or off in the context of a protonmotive force (2.1). As electrical and chemical potential differences may follow a different kinetic pattern, we shall see that the clear separation between chemical and electrical terms, as provided by the TIP procedure, is very helpful. The applicability of the LMA treatment may have a wider range than that of the TIP treatment, but it is also limited to a range in which the highly simplified equations as used in the above treatment are valid.

For these reasons we shall, while dealing subsequently with some special systems, give preference to the treatment in terms of Thermodynamics of Irreversible Processes (TIP).

An experimental test for electrogenicity of a pump can be derived from Eqs. (44a, b and c). After splitting each X_H and X_K into its chemical and electrical term, according to Eqs. (45a and b), we obtain the partial differentials

$$\frac{I}{F}\left(\frac{\partial J_{ch}}{\partial \Delta\Psi}\right)_{\Delta [H^+],\ \Delta [K^+],\ A_{ch}} = (\nu_H - \nu_K)\ L_r \tag{66}$$

and

$$F\left(\frac{\partial \Delta\Psi}{\partial A_{ch}}\right)_{\Delta [H^+],\ \Delta [K^+]} = \frac{(\nu_K - \nu_H)\ L_r}{(\nu_K - \nu_H)^2\ L_r + L_H^u + L_K^u} \tag{67}$$

which obviously are both zero for an electrically silent H^+-K^+ pump ($\nu_K = \nu_K$), but should have finite values for an electrogenic pump ($\nu_K \neq \nu_H$). According to Eq. (66), only with an electrogenic pump should the rate of the metabolic reaction, as for instance determined by the rate of O_2 consumption, glycolysis, ATP hydrolysis or the like at constant concentrations of H^+ and K^+, respond to a change in electric PD. Furthermore, according to Eq. (67), only with an electrogenic pump should the electric PD, if determined at constant $\Delta [H^+]$ and $\Delta [K^+]$, respond to a change in metabolic driving force, A_{ch}, which might be induced by altering the supply of O_2 or of any rate limiting substrate.

At this point, one may add a *third* approach to electrical phenomena, namely that in terms of *electric circuit theory*. As already mentioned, however, this approach is not fundamentally different from that in terms of thermodynamics of irreversible processes; in particular it is also limited to the range in which linear relationships between flows and forces hold with sufficient approximation. Otherwise, the methods mainly differ in their units, which are usually interconvertible by constant factors. The approach in terms of electric circuit has the advantage of dealing with more familiar terms and units, and also of representing complicated relationships by lucid and equivalent circuits (Fig. 8).

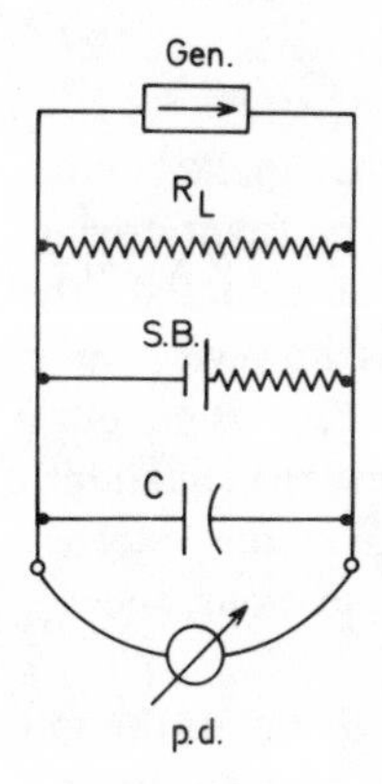

Fig. 8. Electrogenic pump PD in equivalent circuit notation. The electrogenic pump is represented by an electric generator (*Gen.*) functioning as either a constant voltage source or a constant current source. The membrane diffusion PD is represented by a storage battery (*S.B.*) in which electrical energy is partially converted to osmotic energy. The direct contribution to the PD by the electrogenic pump is represented by charging a capacitor (*C*) of comparatively low capacity. R_L is the resistance of the leakage pathways

It becomes less convenient when purely nonelectrical phenomena, forces and flows, are interwoven with electrical ones, as is the rule in most biological systems. This is not the place to discuss the very abundant literature dealing with bioelectric phenomena in terms of equivalent circuits. The reader may be referred for this purpose to the competent and thorough work of W.S. Rehm and his colleagues and disciples, and also to the critical review by Finkelstein and Mauro (1963). The application of the principles of irreversible thermodynamics (TIP) to the problems of electrical networks is thoroughly treated by Kedem and Caplan (1965).

1.4 Membrane Potentials in Secondary Active Transport

1.4.1 Ionic Symporters and Antiporters[1]

In the previous sections only "pumps" are being discussed, with the implication that they are primary active transport systems driven by direct coupling to an exergonic chemical reaction. The same formalism can be applied to secondary active transport of ions, namely to "symport" or "antiport". In analogy to pumps we distinguish also here between electrogenic and electrically silent processes, although for passive flows the terms "rheogenic" and "non rheogenic", respectively are often preferred. Examples for non-rheogenic symport are the Na^+-K^+-Cl^- transport in Ehrlich cells (Geck et al., 1980), the NaCl transport in the gall bladder (Diamond, 1962) and in the intestine (Armstrong et al., 1979), and for non rheogenic antiport, the HCO_3^-.-Cl^- exchange in red blood cells (Rothstein et al., 1976), the Cl^--OH^- exchange in the intestine (Liedke and Hopfer, 1977) and in Ehrlich cells (Heinz et al., 1977) and the Na^+-H^+ exchange in the intestinal and renal brush border membrane (Murer et al., 1976; Kinsella and Aronson, 1980). Rheogenic symport is found in the Na^+, or H^+-linked secondary active transport of metabolites such as sugars and neutral amino acids (Crane, 1977) which will be discussed further below in a special chapter. Rheogenic antiport, such as the ATP-ADP exchange in mitochondria (Klingenberg, 1970) the Na-Ca antiport in the squid axon (Mullins, 1977) and probably the Cl^--$SO_4^=$ exchange in Ehrlich cells (Villereal and Levinson, 1977) should be associated with a primary electrical effect.

The formal treatment of (passive) antiport is quite analogous to that applied before to the exchange pumps, provided that it is based on the assumption of constant driving force, except that the latter has to be set equal to zero. Also here we may proceed either according to LMA or according to TIP. In the absence of active pumps, however, the LMA treatment loses some of its disadvantageous complications, so that there is no longer a reason to prefer the simpler, but less reliable TIP treatment. Otherwise the equations are quite analogous to those developed for the corresponding exchange pumps, except that in view of the conjugate driving force we set $\Gamma = 1$ in the LMA equations, and $A_{ch} = 0$ in the TIP equations.

For *nonrheogenic antiport* the electric PD, as a function of the activities of the solutes involved, can, in analogy to the electrically silent exchange pump, be adequately described by the Goldman-Hodgkin-Katz equation, but using only the rheogenic permeability coefficients for the transient and steady state as well.

1 The terms symport and antiport are used as synonymous with cotransport and counter-transport, respectively. There is a certain tendency to prefer the latter for ion-linked secondary active transport of organic nutrients, such as sugars and amino acids, without any rational justification.

For the *rheogenic antiport* the corresponding procedure is somewhat more complicated; for instance, the transient PD that may appear depends also on the movements of those ions which are not directly involved in the antiport, but are necessary to maintain electroneutrality. Very often, too little is known of such ion movements to allow this procedure. As, however, in passive antiport systems transient states are presumably very short-lived, the simpler steady state equations usually give sufficient information.

To illustrate these points, the steady state equation will be developed briefly for the Na^+/Ca^{2+}-antiport system. Its rheogenicity depends on the electrovalency of the ions (z_i) involved as well as on the stoichiometry (ν_i) of their coupling in the exchange reaction. The antiport process is

$$\nu_{Na}\,Na^{+'} + \nu_{Ca}\,Ca^{2+''} \leftrightarrow \nu_{Na}\,Na^{+''} + \nu_{Ca}\,Ca^{2+'}$$

and its rate

$$J_a = k_a \left[\left([Na^+]'\,\xi^{\frac{z_{Na}}{2}} \right)^{\nu_{Na}} \cdot \left([Ca^{2+}]'\,\xi^{\frac{-z_{Ca}}{2}} \right)^{\nu_{Ca}} - \left([Na^+]''\,\xi^{\frac{-z_{Na}}{2}} \right)^{\nu_{Na}} \cdot \left([Ca^{2+}]'\,\xi^{\frac{z_{Ca}}{2}} \right)^{\nu_{Ca}} \right] \quad (68)$$

J_a is the exchange flow, defined as positive if Na^+ is moved into the cell; k_a is an involved function of mobility and concentrations of free and loaded carrier, respectively, and hence may vary with the degree of saturation. As the antiport is rheogenic, it is restrained by accompanying (leakage) movements required to maintain electroneutrality, i.e., it depends on the electric PD. We may, however, assume that the charge transfer by the antiport is small as compared to the permeance and concentrations of all leaky ions, so that its effect on the distribution of these ions and hence on the electric PD can be neglected.

The total inward movement of Ca^{2+} is

$$J_{Ca} = -\nu_{Ca}\,J_a + J^u_{Ca}\,; \quad (69)$$

it comprises that through the antiport ($-J_a$) and that through leakage (J^u_{Ca}). The negative sign follows from our definition of the sign of J_a on the basis of the Na^+ movement.

$$J^u_{Ca} = P_{Ca}\,[Ca^{2+}]'\,\xi - P_{Ca}\,[Ca^{2+}]''\,\xi^{-1}$$

To the extent that we are interested in the Ca^{2+} distribution at static head, we simply set $J_{Ca} = 0$; implying that now $\nu_{Ca}\,J_a = J^u_{Ca}$:

$$\left(\frac{[Ca^{2+}]''}{[Ca^{2+}]'} \right)^{\nu_{Ca}}_{\text{static head}} = \frac{k_a\,([Na^+]'')^{\nu_{Na}} + P_{Ca}\,\xi^2}{k_a\,([Na^+]')^{\nu_{Na}}\,\xi^2 + P_{Ca}} \quad (70)$$

It appears reasonable to assume that the leakage of Ca^{2+} through the plasma membrane is so slow as compared to the exchange process that it can be neglected. Hence the steady state is sufficiently close to the equilibrium state of the antiport system, which is obtained by setting the exchange flow, J_a, equal to zero. Assuming that $\nu_{Na} = 4$, and $\nu_{Ca} = 1$ (Mullins, 1977) and considering that $z_{Na} = +1$ and $z_{Ca} = +2$, we arrive at Eq. (71) which for practical purposes should give us the final distribution ratio of the Ca-ions maintained by the exchange process as a function of the Na distribution and the electrical potential [Eq. (71)].

$$\frac{[Ca^{2+}]''}{[Ca^{2+}]'} = \left(\frac{[Na^+]''}{[Na^+]'}\right)^4 \xi^{-2} \qquad (71)$$

For normal Na^+ concentrations and PD the equation predicts that at an extracellular $[Ca^{2+}]$ of 2.5 mM, the intracellular $[Ca^{2+}]$ concentration could be as low as 10^{-8} M.

The approach in terms of irreversible thermodynamics for the present system would be as follows:

$$J_a = L_r (\nu_{Na} X_{Na} - \nu_{Ca} X_{Ca})$$
$$= -L_r \left\{ \nu_{Na} RT \Delta \ln [Na^+] - \nu_{Ca} RT \Delta \ln [Ca^{2+}] + (\nu_{Na} - \nu_{Ca}) F\Delta\Psi \right\}$$

Provided that leakage of Ca^{2+} can be neglected, the steady state distribution ratio of Ca^{2+}, as derived from this approach, is

$$\ln \frac{[Ca^{2+}]''}{[Ca^{2+}]'} = \frac{\nu_{Na}}{\nu_{Ca}} \ln \frac{[Na^+]''}{[Na^+]'} + \frac{\nu_{Na} - 2\nu_{Ca}}{\nu_{Ca}} \frac{F\Delta\Psi}{RT}$$

the logarithmic form of Eq. (71).

1.4.2 Ion-linked Cotransport of Organic Solutes

The generation of an electrical potential gradient across a membrane may drive not only the flow of permeant ions but, through an indirect effect, also the flow of nonelectrolytes, to the extent that this is coupled to the flow of ions in "cotransport" (Heinz, 1978). Most extensively studied in this respect is the cation-linked cotransport of sugars and neutral amino acids, which has been recently reviewed by Crane (1977). As this cotransport involves the transfer of a positive charge, it is rheogenic, i.e., the PD is an essential part of the nonconjugate driving force.

Such a function of the electrical PD as a driving force is less obvious if the cotransported solute is itself an ion; for instance, a monovalent anion being cotransported with Na^+ or H^+ at an 1:1 stoichiometry. A typical example is the transport of lactate in Ehrlich cells, which is linked to the flow of H^+ (Spencer and Lehninger, 1976). The "transportate" is neutral so that its net movement across the barrier should not be affected by an electric PD. The question arises whether nontheless the PD should be included in the driving force. Is the latter still the electrochemical PD or only the chemical PD of the co-ion? Does it really make no difference whether the two PD terms were included or omitted? To answer this question we describe the process in "quasi-chemical" notation of TIP. The overall coupled reaction is

$$A^{-'} + H^{+'} \longrightarrow A^{-''} + H^{+''}$$

and its reaction rate is

$$J_r = L_r X_{A^-} + L_r X_{H^+}$$

As the implicit electric terms cancel here

$$X_A + X_H = -RT(\Delta \ln [A^-] + \Delta \ln [H^+])$$

i.e., inclusion or omission of the electric terms makes no difference. It follows that for the mere chemical accumulation of the organic anion, the chemical PD of H^+ sufficiently represents the nonconjugate driving force. If, however, in order to be consistent with thermodynamic usage, we prefer to define the driving affinity as the electrochemical PD of H^+, we have also to express the accumulation of A^- in terms of its electrochemical PD, i.e., to include the electric terms in both the A^- term and the H^+ term. This is also useful in the presence of significant leakage of free ions, which usually also depends on the electric PD. It is seen that owing to the (uncoupled) leakage term, $L_A^u X_A$, the PD terms no longer cancel completely and might for reasons of simplicity be included in the coupled reaction.

In systems of Na^+-linked anion transport in the brush border membrane of intestinal or renal epithelia a positive charge transfer has been observed (Murer and Kinne, 1980) as if *two* monovalent cations, either two Na^+ or $Na^+ + H^+$, were cotransported with each monovalent anion. In that case the "transportate" would no longer be neutral and the coupled flow should depend on the PD according to the following equation, which gives the coupled flow under the assumption that one Na^+ and one H^+ are cotransported with the anion (A^-):

$$J_r = L_r X_A + L_r X_H + L_r X_{Na}$$

$$= -L_r (RT \Delta \ln [A^-] + RT \Delta \ln [H^+] + RT \Delta \ln [Na^+] + F\Delta\Psi)$$

Leakage flows introduce further effects of the PD on the flow of A^-.

In the cotransport of a di- or trivalent anion the possibility might exist that the transportate is an anion, if for instance the number of cotransported cations incompletely neutralizes the anionic charge. Again, the coupled flow becomes dependent on the PD, though in a direction inverse to that of the previous systems, as is seen in the following equation of the coupled flow under the assumption that one divalent anion is cotransported by one Na^+:

$$J_r = -L_r \left\{RT \Delta \ln [A^-] + RT \Delta \ln [Na^+] - F\Delta\Psi\right\}$$

Obviously, the electrical PD that would increase the conjugate driving force for the cation alone counteracts here the translocation of the complex. Can we still maintain that the driving force of this cotransport is the electrochemical PD of the "driver" ion? Does not in this case the driving force seem to be $-(RT\,\Delta\ln[Na^+] - F\Delta\Psi)$, as if the driver ion were an anion? The answer is similar to that in the previously discussed case. As long as we look merely at the (chemical) accumulation of $A^=$, it would indeed be determined by the above expressions, hence *not* by the electrochemical PD of the cation, but seemingly by that of the salt *anion*.

Again, for the sake of thermodynamics consistently we should characterize both the accumulation of $A^=$ and the nonconjugate driving force in terms of electrochem. potential differences, especially in the presence of leakage.

So far, however, it seems that cotransport is hardly ever associated with a negative charge transfer, since no such system has unequivocally been identified yet.

1.5 Redox Potentials in Membrane Transport

Oxidation-reduction reactions are assumed to be directly coupled to proton-translocating systems to form "redox pumps" in the plasma membrane of various microorganisms, in the interior membranes of mitochondria, in the thylakoid membrane of chloroplasts and others. They depend on the presence of electron- and hydrogen-transferring enzymes in such membranes. As the occurrence of such enzymes in the plasma membrane of animal cells is scarce, the function of redox pumps in these membranes is still debated but cannot be excluded (Kilberg and Christensen, 1979).

A special feature of the redox pump is that the driving affinity of the transport, A_{ch}, is usually expressed in terms of an electric PD, the "redox PD". This does not mean that the redox PD can as such always be measured as a membrane PD. It may, however, affect the membrane PD under special circumstances which we shall try to describe in the following.

A suitable model may be based on an H^+ transport loop taken from the "chemiosmotic" system of Mitchell (1966), as depicted in Fig. 9. It consists of two well-stirred compartments separated by a barrier which is permeable only to a hydrogen carrier, SH_2 in the hydrogenated, and S in the dehydrogenated form. At the left side (') S is reduced to SH_2 by a suitable cytochrome-like electron carrier ($Me_1{}^{2+}$) and 2 H^+ according to the equation

$$S' + 2\,Me_1{}^{2+} + 2\,H^{+\prime} \rightleftharpoons SH_2' + 2\,Me_1{}^{3+}$$

SH_2 moves across the barrier to the right side (")

$$SH_2' \rightleftharpoons SH_2''.$$

After being dehydrogenated there by another electron carrier $Me_2{}^{3+}$, according to

$$SH_2'' + 2\,Me_2{}^{3+} = S'' + 2\,Me_2{}^{2+} + 2\,H^{+\prime\prime}$$

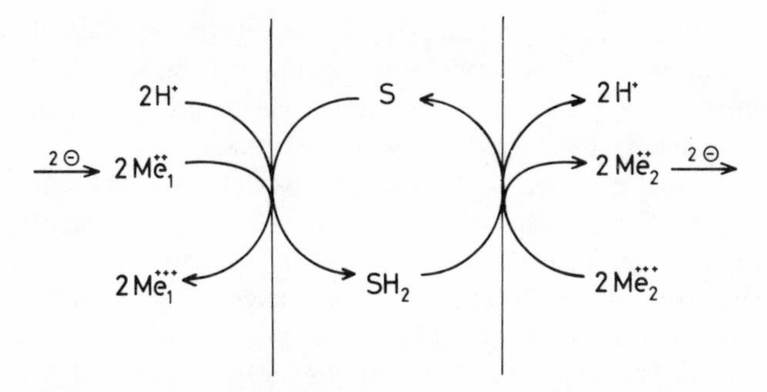

Fig. 9. Redox pump for protons by hydrogen shuttle/single loop. S and SH_2 represented the hydrogen carrier in its oxidized and reduced form, respectively. Me_1 and Me_2 represent cytochrome-like electron carriers, each of them present at one side of the membrane only. To maintain a flow of protons from left to right the two metal systems are connected with an electron donor on the left side and electron acceptor on the right side, e.g., by a pair of shiny platinum electrodes, connected with an electric energy source, and a suitable redox catalyst. A shunting pathway, e.g., for the passive movement of K^+ and H^+, is to maintain electroneutrality. For further details see text

S moves back to the left side

$$S'' \rightarrow S'.$$

The overall reaction, considering that terms referring to components cycling through the barrier drop out, is

$$2\,Me_1^{2+} + 2\,Me_2^{3+} +^{\cdot}\ 2\,H^{+\prime} \longleftrightarrow 2\,Me_1^{3+} + 2\,Me_2^{2+} + 2\,H^{+\prime\prime}$$

and its affinity is

$$A_r = -2\,RT\,\Delta\ln[H^+] + 2\,F\,\Delta\,E_{red}, \text{ in which} \tag{72}$$

$$F\Delta\,E_{red} = F\Delta\,E^0_{red} + RT\ln\frac{[Me_1^{3+}]^2\;[Me_2^{2+}]^2}{[Me_1^{2+}]^2\;[Me_2^{3+}]^2}$$

i.e., the difference between the redox potentials of the two Me systems. As the overall process is electrically neutral, all electric PD terms cancel. For the same reason, the process should proceed toward complete equilibrium without restraint, thereby translocating a limited amount of H^+ from the left to the right side. In order, however, that this H^+ transport can go on continuously, the electrons arriving at the right side must be separated from H^+ and carried back to the left side, making the process electrogenic, as will be discussed below.

Before dealing with such a separation of charges as might come about in the biological system, we shall try to understand the basic principles in a simpler, more mechanistic system in which the required electron flow is provided by a pair of redox electrodes, one on each side, connected with a constant electric electronmotive force (E_{el}), negative on the left, positive on the right side. For reasons of simplicity, we assume that neither

Me electron carrier system, Me_1 and Me_2, can penetrate the barrier, but that each is able to exchange electrons with the electrodes. In this model the electrons arriving at the right side are now forcibly cycled back to the left side through the outside circuit. The maximum driving force for the overall process can be equated with the outside EMF, E_{el}. The question now arises, how much of E_{el} appears as a measurable electrical membrane PD, $\Delta\Psi$, and how much of it is transformed into redox PD (ΔE_{red}).

As mentioned above, the overall cycling process is no longer electroneutral and therefore depends on a parallel conducting pathway (bridge) for charge compensation. As we shall see later, the same is true for the natural system through which the electrons are moved by the respiratory chain. The conducting pathway (bridge) is most likely provided by the leakiness of the barrier to certain ion species, such as K^+ and H^+. In other words, for reason of electroneutrality, the electrons can pass between the electrodes and the corresponding Me systems only to the extent that charge compensation through the mentioned leakage pathways is possible. To the extent that the compensation of charge is prevented from completion, part of E_{el} remains as a measurable membrane PD, $\Delta\Psi$, in parallel to other membrane PD's that may exist. Thus the total E_{el} may appear in two portions: as the redox PD between the two Me systems, ΔE_{red}, which is not measurable, and as the electric membrane PD, $\Delta\Psi$, which is measurable between the two adjacent solutions:

$$E_{el} = \Delta E_{red} + \Delta\Psi$$

In order to introduce this membrane PD, $\Delta\Psi$ into Eq. (72), we simply expand this equation by $+F\Delta\Psi - F\Delta\Psi$, so that we obtain

$$A_r = -2RT\,\Delta \ln [H^+] - 2F\Delta\Psi + 2F\,\Delta E_{red} + 2F\Delta\Psi$$

i.e., $A_r = 2X_H + 2FE_{el}$

$$\text{or} \quad A_r = 2X_H + A_{ch} \tag{73}$$

We see that A_r remains the same whether or not we include the membrane PD. For thermodynamic consistency, however, and also for the sake of greater convenience, especially in the presence of leakage terms, it is recommendable to use electrochemical potential terms, i.e., to include the electric PD. Accordingly the rate of H^+ pumping is

$$J_H = 2L_r A_r + L_H^u X_H$$

L_H^u being the (uncoupled) leakage coefficient, or, inserting A_r from Eq. (73),

$$J_H = (4L_r + L_H^u) X_H + 2L_r A_{ch} \tag{74a}$$

Equation (74a) shows us that at static head ($J_H = 0$)

$$X_H = -\frac{2L_r}{4L_r + L_H^u} A_{ch} \tag{74b}$$

which is maintained by a continuous flow of electrons. It does not, however, tell us how much of X_H is chemical ($-RT\,\Delta \ln [H^+]$) and how much is electric ($-F\Delta\Psi$), because this depends on the relative concentration of the leaking ions in a way that has been discussed previously. To illustrate this we again assume that K-ions stand for all leaking ions. Hence H^+ can be pumped to the right side to the extent that an equivalent amount of K^+ moves from right to left, driven by an electrical potential across the membrane. As soon as the K^+ distribution has reached equilibrium, i.e., its "Nernst" distribution, its net movement and hence also that of H^+ stops. The magnitude of the electric PD at this point depends on the concentration of K^+ relative to that of H^+. If, for instance, the K-ion concentration is very small, a negligibly small amount of K^+ transfer will greatly increase its distribution ratio and hence the electric PD. Accordingly, only very few H-ions can be transported and as a consequence, almost the full outside EMF (E_{el}) appears as a measurable PD ($\Delta\Psi$). On the other hand, if the K^+ concentrations in both compartments greatly exceed those of $H^{\pm}$, considerable amounts of K^+ can be moved from right to left without appreciably changing its distribution ratio. Under these conditions the system can transport a great amount of H^+ before reaching equilibrium with the electrodes, i.e., until the redox PD of the M systems, ΔE_{red} is opposite and equal to the outside EMF, E_{el}. As a consequence, almost nil of E_{el} will appear as a membrane potential ($\Delta\Psi \to 0$). The process is now limited by the chemical PD of the protons, which should come close to E_{el}:

$$RT\,\Delta \ln [H^+] \;\rightarrow\; -E_{el}$$

We see that in the first case the process is limited predominantly by an opposing electrical PD, and in the second extreme case predominantly by an opposing chemical PD of H-ions.

Accordingly, $\Delta\Psi$, the membrane PD, may have values between close to zero in the presence of abundant permeant ions, and close to ΔE_{el} if permeant ions are scarce. At static head, the pump "idles" to the extent that it merely compensates the back leakage of H^+.

Under the condition of constant current maintained through the driving redox electrodes the equations are essentially equal to those derived for the active H^+ pump before [Eq. (56)]. Hence in the transient state, or at stationary flow

$$F\Delta\Psi = \frac{\nu_H J_r}{L_H^u + L_K^u} - \frac{L_H^u}{L_H^u + L_K^u} RT\,\Delta \ln [H^+] - \frac{L_K^u}{L_H^u + L_K^u} RT\,\Delta \ln [K^+] \tag{75}$$

and at static head

$$F\Delta\Psi = -RT\,\Delta \ln [K^+] = \frac{\nu_H J_r}{L_H^u} - RT\,\Delta \ln [H^+] \tag{76}$$

The model on which these considerations and derivations are based, though there are no electrodes or equivalent devices present in a biological system, can essentially be taken to represent a single loop in the middle of the respiratory chain, to which electrons are supplied from one side and removed from the other. If we consider only this single loop, it does not matter where the electrons come from and where they go, as long as this source-and-sink arrangement of electrons is maintained, whether by electrodes or by a pair of electron carrier systems.

If we look at the respiratory chain as a whole we have to consider that the ultimate source of electrons and H^+ is an H-donor (DH_2), e.g., a hydrogenated substrate, and the ultimate sink, a H-acceptor (A), e.g., a dehydrogenated substrate or oxygen (Fig. 10).

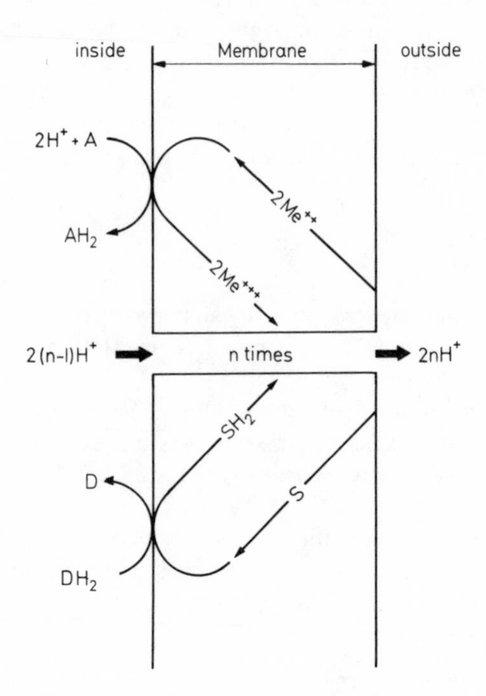

Fig. 10. Redox pump for protons by hydrogen shuttle overall chain. For details see text

The overall reaction, as the terms referring to cycling processes drop out, is

$$DH_2 + A + 2\,n\,H^{+\prime} \rightarrow D + AH_2 + 2\,n\,H^{+\prime\prime}$$

n being the number of energy conserving steps (Mitchell loops) of the whole chain. As the A and D systems are located at the left side only, it is only the translocation of H^+ that makes the overall process electrogenic. The overall affinity is

$$A_r = F(E_D{}^0 - E_A{}^0) + RT \ln \frac{[D]' \cdot [AH_2]'}{[DH_2]' + [A]'} - 2\,n\,RT\,\Delta \ln [H^+] \quad (77)$$

$$- 2\,n\,F\Delta\Psi$$

$$= F(E_D - E_A) + 2\,n\,X_H$$

We see that the overall redox PD, $F(E_D - E_A)$ is not by itself, but only through the H^+-translocating system capable of generating an electric membrane PD. As was the case with the electrode model considered before, the overall redox PD represents the nonconjugate driving force, A_{ch}, of the H^+ pump. At static head

$$A_r = X_H = -RT\,\Delta \ln [H^+] - F\Delta\Psi$$

but the apportioning between the electric and the chemical component of X_H depends on the concentrations and mobilities of the other permeant ions present, in the same way as has been derived for the previous model. A more recent analysis of electrogenic redox pumps in terms of electrical networks has been given by Rehm (1980).

In the systems discussed above the linkage between redox PD and the transport driven by it is primarily stoichiometric, i.e., for each electric charge transferred a fixed number of solute particles is primarily transported. Deviations from this strict stoichiometry are caused only secondarily by leakage. The question arises, whether this is always so, or whether a *nonstoichiometric coupling* between a redox and a transport process is in principal possible. A simple model can theoretically be constructed in which the coupling between a redox process and the movement of a solute is very loose (Fig. 11). To illustrate this we choose again a hydrogen carrier to be involved, occurring in the oxidized (S) and in the reduced (SH_2) form. S could be an aldehyde, and SH_2 the conjugate alcohol. Initially both forms are equally distributed between the two compartments. For the present purpose we make the ad hoc assumption that the membrane is

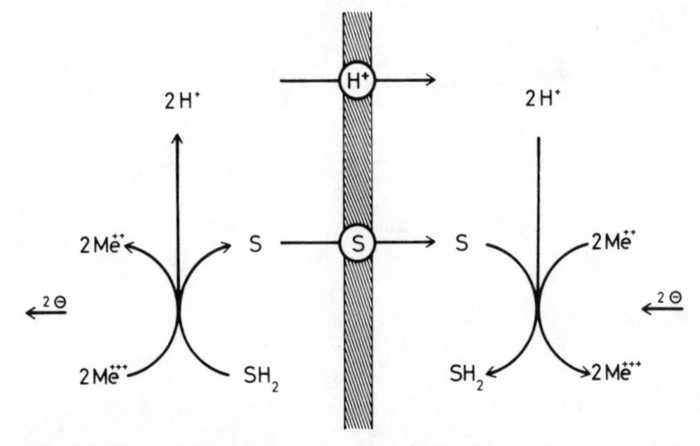

Fig. 11. Nonstoichiometric redox system, leading to the accumulation of electroneutral redox compound (S and SH_2). For details see text

preferentially, or even exclusively, permeable to S, possibly by the operation of a specific facilitated diffusion system. A redox process will now be induced, for example by applying an electrochemical potential from the outside with the help of two platinum electrodes. As such hydrogen carriers (S and SH_2) do not react directly with a platinum electrode, an appropriate catalyst, usually an electron carrier system, is present as a mediator, so that the redox system is not much different from that used in our model for proton transport. At the anode, SH_2 will be oxidized to S, while the opposite will occur at the cathode.

We further assume a proton channel,which not only maintains electroneutrality but also allows for a rapid equilibration between the protons produced on the left side and those consumed on the right side. The proton movement is assumed to be completely separate from the movement of S, and to be rapid enough not to limit the overall rate. The oxydation-reduction process now taking place will put both S and SH_2 into disequilibrium, but according to our assumption only S can escape via the membrane. As a consequence, S will steadily move from left to right, following its concentration gradient, and the total amount of redox substrate (S and SH_2) will increase in the right compartment.

It is clear that the electrical current passing through this system is coupled to an electrically silent process, namely to the movement of a solute which is neither electrically charged nor directly linked to a charged particle. This coupling is indeed not primarily stoichiometric, as the rate of solute transport depends on the selective permeability of the membrane for this solute, which may be completely different from the permeability for protons. Only in the final steady state, which may be reached after a long time, does the overall reaction become seemingly stoichiometric, in that for each electron pair reacting one S will move to the opposite side, provided that the membrane is completely impermeable to SH_2.

Perhaps comparable to such a hypothetical system of (nonstoichiometric) redox-PD-driven transport is the accumulation of xylitol in the lens of the eye, which appears to be the basis of a xylose-induced cataract (van Heyningen, 1959). It is assumed that the membrane of the lens concerned is less permeable to xylitol than to the conjugate aldehyde from which it is formed by a reductase.

2 Control of Electric Potentials – Maintenance and Modulation

2.1 Physiological Mechanisms

A stationary PD of a certain magnitude appears to be required for normal metabolic activity of the cell, and presumably also for "normal" permeability of the cell membrane to various solutes. Accordingly, perturbations of the PD are usually associated with alterations of cellular functions, in particular of the plasma membrane (Katz, 1966; Harold, 1977). It is often difficult, however, to tell which comes first, i.e., whether a change in PD causes a change in permeability or vice versa, or whether both changes are coinciding results of some other change, for instance in the concentration of certain ions. It is also possible that permeability changes that occur with changes in PD may in reality result from alterations in the static phase boundary PD at the membrane surfaces, even though the latter are not directly related to membrane permeability.

Changes of the electrical PD of biological membranes, in particular of cell membranes, are presumably of great importance for the operation and regulation of vital functions. This has long been known for excitable tissues such as muscle cells, and nerve cells and fibers (Katz, 1966), and there is good reason to assume that it applies to cells and cellular organelles in general.

As has been mentioned before, two major biological functions are attributed to potential changes: first, to serve as a signal to trigger another process, such as excitation of nerves and muscles; second, to provide a prompt and powerful source of energy to drive highly endergonic processes, such as ATP synthesis in mitochondrial and bacterial membranes, or the ion-linked active transport of amino acids and sugars and other solutes into and through the cell.

It seems that these two functions correspond to the two different physiological mechanisms of a PD change: (1) a change in the rheogenic permeability for a distinct ion species; (2) a change in the power of an electrogenic pump. The first of these mechanisms appears to account for the signal effects and the second for the supply of energy.

The action potential may serve as an example of the first kind: this is a rapid and drastic change of electrical membrane diffusion potential resulting from a sudden modulation of the membrane permeability for a distinct ion species. The action PD is involved in the production (triggering) and propagation of an impulse in excitable tissues, such as nerve cells, nerve fibers, muscle cells, and other cells, where it may initiate special physiological functions, e.g., the contraction of the muscle fiber. In the present context we can discuss these physiological processes only to the extent that the principles discussed in previous chapters can be directly applied. For more detailed information the reader may be referred to modern textbooks of physiology, or to special treatises (Katz, 1966; Finkelstein and Mauro, 1977).

The sequence of events associated with the generation of an action potential has best been elucidated for the giant squid axon. In its physiological environment the axon normally contains an intracellular fluid with a high K^+ and a low Na^+ concentration, and is surrounded by extracellular fluid, which has a high Na^+ and a low K^+ concentration. In the resting state, an electric PD of about −120 mV, inside negative, can be measured across the axon membrane. This "resting PD" can be interpreted in terms of a membrane diffusion potential, mainly due to the unequal distributions and permeances of K^+ and Na^+. In the resting state, this membrane potential appears to be dominated by the K^+ distribution, as it comes close to the so-called "Nernst PD" of the K-ion. Upon initiation of a nerve impulse, the membrane PD is "activated", i.e., it rapidly drops below the zero line toward a peak value (spike) of about +40 mV (inside positive) and subsequently just as rapidly returns toward its original value, temporarily even somewhat below it ("undershoot"). The PD at the spike comes close to the Nernst PD of the Na-ions (+50 mV), which appears to indicate that it is temporarily dominated by the Na^+ distribution. Accordingly it has been demonstrated that this inversion of the PD is indeed associated with a sudden and transient 40-fold increase in membrane permeability for Na^+.

The appropriate stimulus for the sudden change in Na^+ permeability appears to be a lowering of the membrane PD (depolarization), which after it has fallen below a threshold value of about −50 mV appears to precipitate the increase in Na^+ permeability required for the action PD. This illustrates how the specific permeability of a membrane may be affected by a PD change, here in terms of a positive feedback loop: the initial change of the PD prompts a further change of this PD in the same direction.

It was also shown that the decline of the action PD is brought about by a change in permeability: the permeability of the membrane for Na-ions drops to its resting value just as rapidly as it has been raised before (Fig. 12). Apparently the positive feedback mechanism by which the initial depolarization opens the Na^+ channel is suddenly interrupted (inactivation) after the peak value has been reached; the Na channel returns to, and remains in, its original state of low permeability. Also this process can be formally explained in terms of an appropriate equation which relates to PD of specific permeabilities of the membrane to certain ion species. The whole process is complicated by the fact that the K^+ permeability of the membrane also starts rising soon after the initiation of the impulse, but the increase in K^+ permeability is smaller and develops much more slowly than that of the Na permeability. So the increase in K^+ permeability is hardly effective up to the peak of the potential, but it may already be significant when the action PD is on its way back, i.e., after the Na^+ permeability has already returned to its resting value. The increase K^+ permeability is supposed to accelerate the reversal of the action PD and is presumably also responsible for the previously mentioned undershoot (Fig. 12).

So far all electrical events in connection with action potential in excitable tissues can be quantitatively accounted for by the quick and specific permeability changes in the cell membranes, in line with the equations previously developed for the relationship between electrical potential on one hand and permeabilities and concentrations of special ions on the other hand. Both approaches, that based on LMA [Goldman-Hodgkin-Katz-equation (23)], and that based on non-equilibrium thermodynamics are capable of

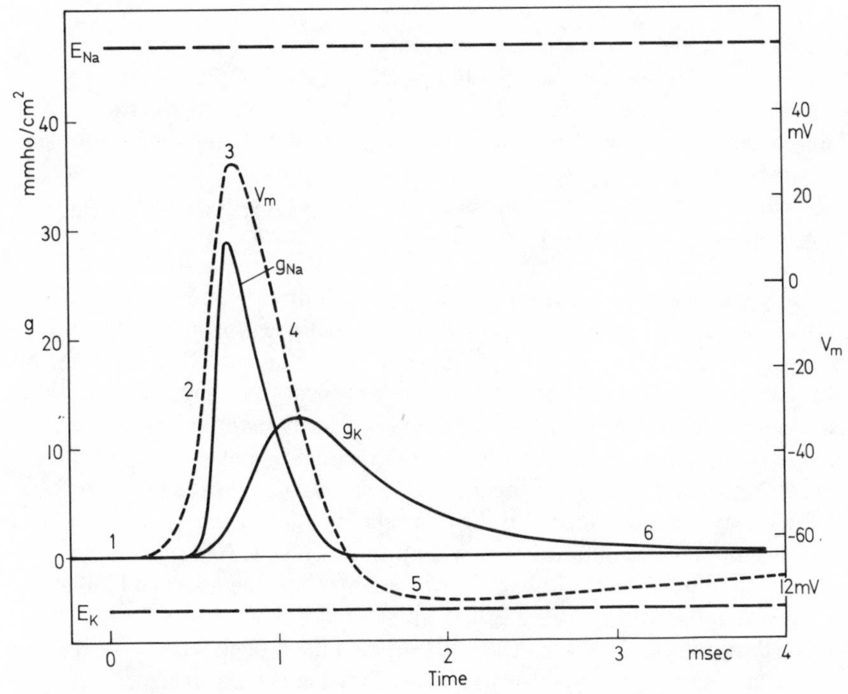

Fig. 12. Action potential as a function of changes in membrane permeabilities. The curves are obtained by solving the equation derived by Hodgkin and Huxley for the giant axon of the squid at 18.5°C. The *solid lines* represent the membrane conductances g_A and g_K of Na^+ and K^+, respectively. The *dotted lines* represent the resulting membrane PD (V_m). E_{Na} and E_K represent the equilibrium (Nernst) PD's for Na^+ and K^+, respectively. *Abscissa:* time in milliseconds (ms); *ordinate left:* conductance in S/cm^2; *ordinate right:* membrane PD in mV

explaining the observations on the basis of the permeability changes described. The mechanistic basis of these permeability changes is attributed to the opening (or closing) of "gates" of ion-specific channels, triggered by the depolarization event (Keynes, 1977).

The protonmotive force (PMF), controlled by turning on or off an electrogenic proton pump, physiologically or by experimental manipulation, may illustrate the other kind of potential-generating mechanism. As a practical example we may choose the proton pump in mitochondria and certain bacteria, which has been treated quantitatively earlier in this booklet (1.3). This pump tends to push H^+-ions across the barrier but can achieve net transport only if H^+ and other permeant ions, e.g. K^+, move passively to maintain electroneutrality. As however these passive ion currents are presumably sluggish as compared to the pumping rate, the pump will first charge the static capacity of the membrane, thereby causing a sharp rise of the potential, possibly followed by a slower PD decline while a chemical PD of H^+ is gradually built up. To the extent that the passive current is carried by protons, it has no consequence for the system as it merely restores the protons to the compartment from which they originate. The potassium current, however, will cause the accumulation of K^+ in the one compartment and of H^+ in the other, thus loading a "concentration cell" consisting of two opposing diffusion PD's, of K^+ and H^+, respectively, in which some of the pumping energy is being conserved. The whole process will proceed until static head is reached, at which net transport of all ions ceases, and at which K^+ is in complete equilibrium, the potential of the concentration cell being identical with the Nernst PD of K^+. At this stage, however, the electrochemical potential difference of H^+ is maintained at a value high enough to make the protons leak back at the same rate as they are pumped.

The time course of these events is complicated by the fact the electric and the chemical component of the protonmotive force follow different kinetic patterns. It has been treated mathematically for the protonmotive force during an oxygen pulse in anoxic mitochondria by Mitchell (1968) under the assumption of constant pumping rate. The result was that the electric PD indeed rises much more rapidly, within about 100 ms, than does the pH-difference, which has a half time of about a minute. The magnitude of the early PD rise, however, turned out to be too small to be of any significance. Meanwhile it has been found out under another assumption, namely that the driving force rather than the rate of the pump remains constant during the O_2-pulse, that the initial PD rise is much higher and may even come close to the maximum protonmotive force reached in the steady state, provided that the rate coefficient of the pumping force sufficiently exceeds the leakage coefficients, as will be derived below. Hence under such conditions the proton pump might be capable of "anticipating" the maximal PMF at an early stage by a rapid rise of the electric PD (Heinz, 1981).

Extending these considerations to the event following the sudden termination of the pump gives similar results in the opposite direction: a sudden reversal of the electric PD within less than 100 ms and a much slower dissipation of the ion gradients (Fig. 13).

In the present context we shall present only the basic differential equations applicable to the above model, using as an example an electrogenic proton pump with a membrane which is passively permeable only to protons and K-ions.

The change of X_H is the sum of the changes of its components:

$$\frac{d X_H}{dt} = -\left[\frac{d(RT\,\Delta \ln [H^+])}{dt} + \frac{d(F\Delta\Psi)}{dt}\right] \tag{78}$$

$$\frac{d(RT\,\Delta \ln [H^+])}{dt} = \frac{J_H}{B} \tag{79a}$$

and

$$\frac{d(F\Delta\Psi)}{dt} = \frac{J_H + J_K}{C} \tag{79b}$$

B is the effective buffer capacity, a function of the buffer capacities of the solutions in the two compartments,

$$\frac{1}{B} = \frac{1}{B'} + \frac{1}{B''}$$

and C the electric capacity of the membrane. Both B and C refer to the same standard unit, e.g., of weight of mitochondrial protein present.

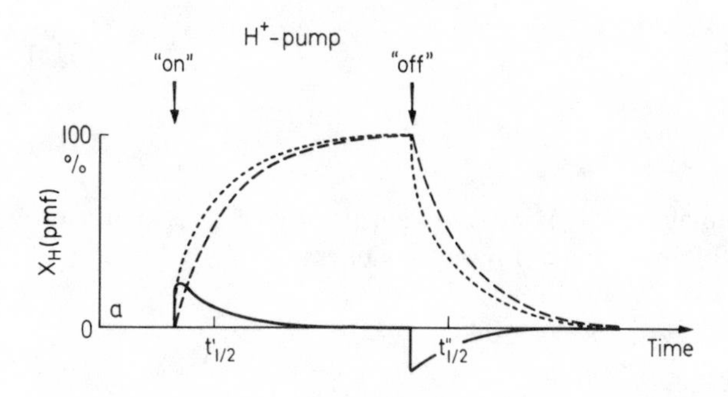

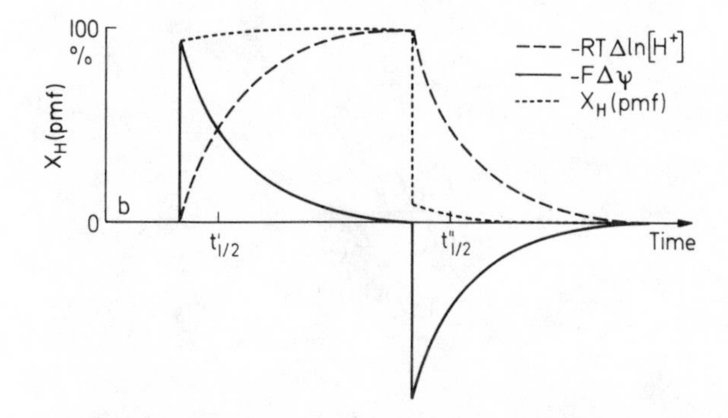

Fig. 13a, b. The response of the protonmotive force to a "pulse" of an electrogenic H^+ pump. The time course of the protonmotive force (PMF) (*dotted line*) and its components, the electrical PD (*solid line*) and the chemical PD (*broken line*) in response to a square wave-like pulse of H^+ pump according to Eqs. (81a and b). ***Abscissa:*** time. ***Ordinate:*** X_H, the "protonmotive force" and its components in percent of maximum (steady state) protonmotive force generated by the electrogenic pump. **a** Under the assumption of constant pumping rate during pulse. The rate coefficients of the passive ion movements (leakage) of H^+ and K^+, respectively, for the buffer capacity (B) and for the electrical capacity (C) are the same as those used by Mitchell (1968). **b** Under the assumption of constant driving force. The magnitudes of L_H, L_K, B, and C are chosen similar to those under **a**. L_r, the overall pump coefficient, is assumed to be 3 times higher than the leakage coefficient for K^+. In this model, the amplitude of the electric peak is rather independent of the ratio of $\frac{L_K}{L_H}$ as long as L_r is sufficiently greater than L_K. Under the conditions chosen, more than 90% of the electrical peak value should be reached after far less than 100 ms after the start of the pump

Hence

$$\frac{dX_H}{dt} = -\left[\left(\frac{1}{B} + \frac{1}{C}\right) J_H + \frac{1}{C} J_K\right] \tag{80}$$

The integration is somewhat complicated because the electrical and the chemical PD each have different kinetics. Under certain conditions, especially if the concentration of K^+ is so high that changes in chemical PD of K^+ can be neglected, the following equations give an approximate description of the two processes

$$RT\,\Delta \ln [\mathrm{H^+}] = -A_1\,(1 - e^{-\lambda_1 t}) \tag{81a}$$

$$F\Delta\Psi = -A_2\,(e^{-\lambda_1 t} - e^{-\lambda_2 t}) \tag{81b}$$

Depending on whether the pumping rate (J_p) or the driving affinity (A_{ch}) is kept constant, the parameters have the following values for our model:

	A_{ch} = constant	J_p = constant
$A_1 =$	$\dfrac{\nu_H L_r A_{ch}}{\nu_H^2 L_r + L_H^u}$	$\dfrac{\nu_H J_p}{L_H^u}$
$A_2 =$	$\dfrac{\nu_H L_r A_{ch}}{\nu_H^2 L_r + L_H^u + L_K^u}$	$\dfrac{\nu_H J_p}{L_H^u + L_K^u}$
$\lambda_1 =$	$\dfrac{(\nu_H^2 L_r + L_H^u)\,L_K}{B\,(\nu_H^2 L_r + L_H^u + L_K^u)}$	$\dfrac{L_H^u L_K^u}{B\,(L_H^u + L_K^u)}$
$\lambda_2 =$	$\dfrac{\nu_H^2 L_r + L_H^u + L_K^u}{C}$	$\dfrac{L_H^u + L_K^u}{C}$

It is seen that with a constant A_{ch}, provided that $B \gg C$ and $L_r > L_K + L_H$, the electric PD rises very fast to a high peak and subsequently declines[1], whereas the chemical PD rises more slowly toward the static head value. In this way much of the pumping power already becomes available at an early stage, long before appreciable ion movements have taken place. It is also seen that this is much less pronounced at constant J_p: the electric PD also rises prior to the chemical PD, but not as fast, and the peak reached is probably very small and transient, so that the pumping power becomes fully available only after the chemical PD of H^+ has been built up (Heinz, 1981) (Fig. 13). At constant A_{ch} an electrogenic pump would have an obvious advantage over an electrically silent one in being able to provide the almost maximal protonmotive force for useful purposes at a very early stage after initiation of the pump, presumably well in advance of the generation of a chemical potential gradient. This is likely to apply to the proton pumps in mitochondria (Mitchell, 1966; Racker, 1978), chloroplasts (Junge, 1975), halophilic bacteria (Stoeckenius et al., 1979) and other microorganisms, and in reconstituted systems (Westerhoff et al., 1979).

The electrogenic proton pump may be taken as a model of other electrogenic pump systems functioning in the biological realm which may be turned on or turned off by hormonal or other regulatory mechanisms. Examples are the activation of the Na^+ pump

---

1 It should be pointed out that the complete decline (and replacement by the chemical PD) of the electric PD, as depicted in Fig. 13, follows from the simplifying assumption that K^+ is in great excess of (free and buffered) H^+, so that the net movement of the latter is of negligible effect on the K^+ distribution ratio. In reality, the electric PD may only disappear in part, especially as the mitochondrial buffer capacity is not negligible and seemingly expanded by a K^+/H^+ antiporter assumed to exist in the mitochondrial membrane

in fibroblasts (Villereal and Cook, 1978), in muscle fibers (Mullins and Awad, 1965), in nerve tissue (Ritchie, 1971), in Ehrlich cells (Heinz et al., 1975), in neurospora (Slayman et al., 1973), and in other tissues.

2.2 Experimental Modulations of PD

To calibrate methods of PD measurements, or to study the biological effects of PD changes, it is desirable to induce or prevent such changes under controlled conditions at will. Of the several methods to manipulate the membrane which have so far been developed and used for artificial and biological membranes, a few will be briefly discussed here.

2.2.1 Replacement of Permeant Ion Species

Replacing these by others with different permeability properties should change the PD in accordance with the Goldman-Hodgkin-Katz equation [Eq. (23)]. If, for instance, the extracellular Cl^--ions of a cell suspension are replaced by a nonpermeant ion species like toluoldisulfate or mucate (Vidaver, 1964a, b) one can expect that the membrane PD (inside negative) becomes smaller (depolarization) (Katz, 1966). This would also occur in the presence of an electrogenic pump, for instance of Na^+. More Na^+ would be pumped out of the cell after the replacement until a new steady state is attained which is characterized by an increased chemical potential difference of Na^+ and, concomitantly, a decreased electric PD. The opposite would occur if we replaced Cl^- by a more permeant ion species, for instance by SCN^- or benzoate (Murer and Kinne, 1980). The Goldman-Hodgkin-Katz equation would now predict an increase in PD (hyperpolarization). An electrogenic outward pump of Na^+ would no longer be able to maintain the previous Na^+ distribution: hence in the new steady state the chemical PD of Na^+ would be lower and the electrical PD higher than previously.

An analogous, though converse, change could be expected if by appropriate pretreatment the replacing ion species were maneuvered into the cell. Replacing intracellular Cl^- by a less permeant anion would increase, and by a more permeant anion, decrease the electrical PD. All these effects are more or less reduced ("swamped out") by the presence of other permeant ions in the system, depending on their concentrations and permeability coefficients (P_i).

Since these effects involve substantial net movement of electrolytes between the cell and medium, they are presumably associated with osmotic water shifts, in other words, with shrinking or swelling of the cell.

2.2.2 Ionophores

An elegant and widely used means to modify cellular membrane potentials is the application of ionophores. These are lipid soluble macromolecules which, if present in even very small amounts, strongly and more or less selectively increase the permeability of the membrane to distinct ion species, thereby modifying the electrical PD across these membranes. Since they act similarly on biological and on certain artificial membranes, the latter can be used as models to study the mechanism of their effect in general (Pressman, 1968).

2.2.2.1 Carrier-like and Channel-like Ionophores

As for the mechanism of their action one distinguishes between carrier-like and channel-like ionophores (Ovshinnikov and Ivanov, 1977) (Fig. 14). The *carrier-like ionophores*

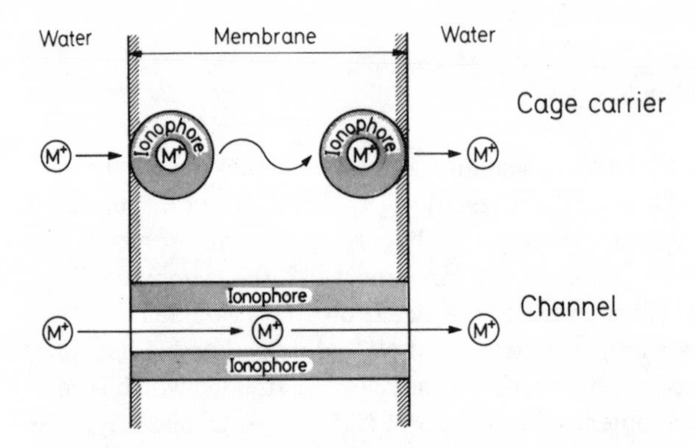

Fig. 14. Carrier (cage)-like and channel-like ionophores and their presumable function in the membrane. (Y.A. Ovshinnikov, V.T. Ivanov in: Biochemistry of Membrane Transport, *42nd FEBS Symposium*, eds. G. Semenza, E. Carafoli. Berlin-Heidelberg-New York: Springer, 1977)

are assumed to act by forming highly permeant lipid-soluble complexes (cages) with the ion species concerned (Harold et al., 1974; Pressman and deGuzman, 1975; Stark, 1978). They have often cyclic structures which allow specific binding of the ion in their center. The best-known example is the K-ionophore valinomycin discovered by Moore and Pressman (1964). The *channel-like ionophores* are assumed to transmit ions by the formation of transient, more or less ion-specific channels through the membrane (Urry et al., 1975; Hall, 1978). They are usually linear polypeptides tending to form tube-like structures which appear and disappear in rapid succession, as if each channel were oscillating between two conformational states, an open one and a closed one. Typical examples are gramicidin and alamethicin. They differ from carrier-like ionophores by, among other things, the insensitivity of their conducting properties to a lowering of temperature, which by increasing the viscosity of the lipid phase of the membrane severely impedes the action of carrier-like ionophores. Furthermore, channels seem to be less specific with respect to the transmitted ion species than are carriers, but may permit a much more rapid penetration of the conducted ion (Mullins, 1975).

2.2.2.2 Rheogenic and Non-Rheogenic Ionophores

As for the behavior of ionophores in the electrical field, one distinguishes between rheogenic and nonrheogenic, or electrically silent ones. The *rheogenic ionophores* permit the net passage of ionic charges through the membrane, thereby conducting an electric current. For this effect it makes little difference whether or not they have an electric charge of their own in the unloaded (empty) state. For instance, a neutral ionophore takes up the charge of the ion it binds, as the neutral valinomycin and monactin become positive after binding K-ions. Hence, only the translocation step, i.e., of the loaded carrier depends on an electric field. If, on the other hand, the ionophore itself is an anion that happens to be neutralized by the cation it binds, then only the relocation step, i.e., that of the empty carrier should be affected by an electric field. Examples are to be found among proton conductors which are likely to be weak acids and hence anions in the unloaded state but become neutral by binding a proton, as has been demonstrated for carbonylcyanide chlorophenyl hydrazone (CCCP) by Le Blanc (1971). No matter whether the loaded ionophore or the unloaded one carries the charge, the direction of the electrical field effect on the overall transport should depend only on the charge of the transportee. Kinetically, however, there might be considerable differences between the two alternatives, especially if the loaded and unloaded carrier differ in their mobility (Heinz and Geck, 1978).

The *nonrheogenic ionophores* act by exchanging electrically equivalent ions so that with each charge moving in the forward direction the same charge moves in the (reverse) backward direction. Usually these ionophores are nonpermeant ions if unloaded but become permeant if neutralized by loading with ions. Typical examples are nigericin, which facilitates an electroneutral exchange between K^+ and H^+, and monensin which acts similarly but prefers Na^+ to K^+ in the exchange for protons (Table 1).

The *channel-like ionophores*, to the extent that they are so "rigid" that neither an electric field nor the passage of an ion may dislocate the postulated binding sites, should be rheogenic. This appears to apply to those ionophores so far studied with artificial membranes whose ionophoric action is inferred from their contribution to membrane conductance (Hall, 1978). On the other hand, certain "gated" channels, such as supposedly occur in biological systems, might function nonrheogenically, since the "opening" and "closing" of such gates could be imagined to require that two ions pass in opposite directions, as has been postulated for the Cl^--HCO_3^--exchange mechanism in human red blood cells (Rothstein et al., 1976).

Whether there are artificial compounds that act as such gated channels if experimentally added to a membrane is not known yet.

2.2.2.3 The Electric PD Under the Influence of Ionophores

Both rheogenic and nonrheogenic ionophores are capable of *altering the electric PD*, the former through a change in permeability of ions, the latter through a change in distribution of ions. The *rheogenic ionophores* are usually preferred for this purpose because they affect the PD directly with little or no change in ion concentrations. Their effect will be illustrated by typical examples: K^+-conductors (valinomycin) and proton conductors.

Table 1. Ionophores

	Selectivity	Reference
I. Rheogenic Carriers		
Valinomycin	K^+	Moore and Pressman (1964)
Nactin group (nonactin, monactin, diuactin)	K^+	Graven et al. (1967)
Enniatin B	K^+	Simon et al. (1969)
2:4 Dinitrophenol	H^+	Mitchell and Moyle (1967)
Carbonylcyanide-phenylhydrozones (CCCP, FCCP, CFCCP)	H^+	Heytler and Pritchard (1962) Le Blanc (1971)
Tetrachlorosalicylamilide	H^+	Harold et al. (1970)
Tetrachloro-trifluoromethyl benzimidazol	H^+	Skulachev (1971)
Di (pentafluorophenyl) mercury	anions (unspecific)	Skulachev (1971) Andersen et al. (1976)
II. Nonrheogenic (antiporter) Carriers		
Nigericin	K^+/H^+	Graven et al. (1966)
Monensin	Na^+/H^+	Stark (1969); Simon et al. (1969); Henderson et al. (1969)
A 23187	$Ca^{2+}H^+$	Reed and Lardy (1972)
Organo-tin compounds (triethyl-, tripropyl-tin)	Cl^-/OH^-	Selwyn et al. (1970)
III. Channel-forming Ionophores		
Gramicidin	cations	Chappell and Crofts (1965)
Alamethicin	(unspecific)	Mueller and Rudin (1969)
Monazomycin	(unspecific)	Mueller and Rudin (1969)
Amphotericin B (?)	anions	Cass et al. (1970)
Nystatin (?)	(unspecific)	Cass et al. (1970)

We use again one of the models discussed before: two compartments, separated by a rigid membrane, containing the permeant ions H^+, K^+, and Cl^- at different concentrations in each compartment. The movement of solvent is assumed to be prevented by the rigidity of walls and membrane. The electrical potential can be approximated by the Goldman-Hodgkin-Katz equation as follows:

$$F\Delta\Psi = -RT \ln \frac{P_K [K^+]'' + P_H [H^+]'' + P_{Cl} [Cl^-]'}{P_K [K^+]' + P_H [H^+]' + P_{Cl} [Cl^-]''}$$

The effect of a K^+-conductor is to provide an additional rheogenic pathway for K^+, which amounts to increasing P_K, the effective permeability coefficient of K^+, by possibly several orders of magnitude. As a consequence the PD will change, depending on the distribution of K^+. If the terms containing P_K become so great that the other terms can be neglected, the potential will come close to the equilibrium (Nernst) potential of K^+. It should be kept in mind, though, that this equilibrium is not yet complete, so that the ions will slowly move toward the true equilibrium and the electrical potential will concomitantly decay.

If we had instead added a proton conductor to the system, the PD would change by the same mechanism, but in the opposite direction, as in our model $[H^+]' > [H^+]''$.

These ionophores will change the PD only to the extent that the accelerated ion species is not in equilibrium with this PD. If, for instance, the K-ions are distributed according to the Donnan equilibrium, the addition of valinomycin should not change the PD.

Electroneutral ionophores are less suitable for the purpose of changing the electrical potential. As has already been discussed, the ionophoric facilitation of the exchange between two strictly electrically equivalent ions should primarily not affect the potential. Still, a change in electrical potential may result as a secondary effect, as such an ionophore, depending on the relative concentrations of the ions concerned, may alter the distribution of the one species more than that of the other, and thereby alter the PD. This can easily be illustrated with the same model as before.

If instead of valinomycin or a proton conductor, we added nigericin to the system, the K-ions of the one compartment would rapidly exchange with the H-ions of the other until the distribution ratio (r_1) between the two chambers was the same for both ions. The Goldman-Hodgkin-Katz equation would then be

$$F\Delta\Psi = -RT \ln \frac{r_1\,(P_K\,[K^+]' + P_H\,[H^+]') + P_{Cl}\,[Cl^-]'}{(P_K\,[K^+]' + P_H\,[H^+]') + P_{Cl}\,[Cl^-]''} \tag{82}$$

If the Cl^- ions were completely impermeant this PD would equal the Nernst PD of each ion:

$$F\Delta\Psi \approx -\frac{RT}{zF} \ln \frac{[H^+]''}{[H^+]'} = -\frac{RT}{zF} \ln \frac{[K^+]''}{[K^+]'} = -\frac{RT}{zF} \ln r_1 \tag{83}$$

which no longer depends on the rheogenic permeabilities of H^+ and K^+. It would most likely be different from the PD before the ionophoric action. If the Cl^--ions were permeant, this PD would be smaller than $-RT \ln r_1$, and also transient since the distribution of the Cl^--ions would tend to adjust to the PD until a final equilibrium were reached at which

$$\frac{[K^+]''}{[K^+]'} = \frac{[H^+]''}{[H^+]'} = \frac{[Cl^-]'}{[Cl^-]''} = r_2$$

so that the final PD would be

$$F\Delta\Psi = -\frac{RT}{zF} \ln r_2 \tag{83}$$

Clearly the final PD is smaller than that with r_1. If all ions present are permeant, the unequal distribution of H- and K-ions will eventually disappear and so will the electrical potential difference.

The direction of the change in PD, i.e., whether the addition of the nonrheogenic ionophore will tend to increase or decrease the PD, depends on the rheogenic mobilities and the changes in activity of the exchanging ions, here K^+ and H^+, only. Unless the rheogenic permeability of H^+ is very much greater than that of K^+, as appears unlikely in most biological membranes, the change in PD will probably follow the change in distribution of K^+, because owing to the buffering capacity of biological fluids the H^+ activities will remain small, in spite of substantial shifts of H^+ through the membrane.

Many ionophores, in the concentrations applied, may additionally inhibit metabolism. As a consequence, the function of active ion pumps will in most cases be depressed, if not entirely blocked. In the end only the PD predictable from the Goldman-Hodgkin-Katz equation remains, which may be very different from the previous one influenced by the electrogenic pump.

2.2.3 Channel Blockers

An alternative means to vary a membrane PD should be by specifically plugging up existing pathways rather than by opening new ones. Substances to serve this purpose, with sufficient specificity, are available, though with smaller variety. Some of the best known are tetrodotoxin (TTX), which acts by obstructing specifically the Na^+ channels in nerve fiber membranes without much affecting the permeability of K-ions, and tetraethyl ammonium or diamino pyridine ions, which analogously affect the K^+ channels of the same membranes (Table 2). A direct and drastic effect of these on the membrane PD is to be expected only to the extent that they obstruct rheogenic pathways. Substances which act on nonrheogenic co- or counter-transport mechanisms will have no or only indirect

Table 2. Permeability blockers

	Selectivity	Reference
Cations		
Tetrodotoxin (TTX)	Na^+	Hille (1966)
Saxitoxin	Na^+	Ritchie and Rogart (1977)
Tetraethyl ammonium (TEA)	K^+	Armstrong and Binstock (1965)
3,4-diamino pyridine	K^+	Kirsch and Narahashi (1978)
Amiloride	Na^+	Bentley (1968)
Verapamil	Ca^{2+}	Bayer et al. (1975)
D-600	Ca^{2+}	Bayer et al. (1975)
Anions		
Stilbene disulfonate derivatives	Cl^-/base (unspecific)	
(SITS)		Cabantchik and Rothstein (1972)
(DIDS)		Cabantchik and Rothstein (1974)
Local anesthetics (dibucaine)	Cl^-/base (unspecific)	Gunn and Cooper (1975)
Probenecid (benemid)	Cl^-/base (unspecific)	Berner and Kinne (1976)

effects on the PD. This may apply to the disulfonic stilbene derivates (DIDS and SITS) widely used to block the Cl^--HCO_3^- exchange in erythrocytes or to furosemide and analogues (Table 2), which presumably block the nonrheogenic cotransport of Na^+, K^+, and Cl^- (Geck et al., 1980).

The electrotropic effect of the specific blockage of a rheogenic ion channel can be predicted on the basis of the Goldman-Hodgkin-Katz equation, considering that these blockers do just the opposite of what a corresponding ionophore would do.

2.2.4 Modulating and Mimicking Electrogenic Pumps

In the introduction it was mentioned that a membrane PD may be altered rapidly and thoroughly by a sudden change in the power of an electrogenic ion pump. Whenever such pumps are naturally present in a system under investigation it should be possible arbitrarily to effect PD changes by stimulating or inhibiting such a pump. For some pumps specific moderators are known, e.g., the electrogenic Na-K pump in many cells is specifically and rapidly stimulated by raising the extracellular K^+ activity, or can be equally rapidly inhibited by cardiac glycosides such as ouabain. The proton pump of the gastric mucosa is specifically stimulated by histamine and inhibited by SCN^-. There is of course always the possibility of modifying an electrogenic pump nonspecifically, e.g., via cellular metabolism, but these methods are less useful because of undesired effects on other metabolism-dependent processes. There is at least theoretically the possibility of mimicking electrogenic pumps by ion-specific electrodes in connection with an outside electric source. Not too many electrodes suitable for this purpose are available, and their application to biological systems appears to be more complicated than that to artificial membranes. On the other hand, to the extent that the electric parameters of such electrodes are clearly understood they may serve as theoretical models to study the effect of electrogenic pumps and as a basis to derive useful equations.

3 Some Problems Associated with the Measurement of Electric Membrane Potentials

3.1 General

To determine absolute electrical potentials across biological membranes precisely and reliably is very difficult. In many cases one may at best obtain approximative values based on assumptions which cannot be tested rigorously. From the foregoing, on the other hand, it is clear that some at least approximative information about electrical potential is extremely important for the study of membrane functions. Hence considerable efforts have been made to improve the technique in order to keep unavoidable errors small enough not do endanger major conclusions drawn concerning the system under investigation. The best way to obtain fairly trustworthy results at the present time may be to apply more than one independent method, so that the values of both may confirm each other. In the following the most important methods presently available to estimate membrane potentials are listed according to the underlying principle. Special emphasis is given to those methods applicable to single cells and subcellular organelles and vesicles. For more information, see the review by Rottenberg (1975).

3.2 Microelectrodes

Electrodes used to measure an electrical potential between two compartments, either for its own sake or as an indirect means to measure the difference between (chemical) activities of a given ion between the two compartments, can be divided into reversible and irreversible types (Fig. 15).

Ideally *reversible electrodes* are in direct contact with the test solution and respond directly to the electrochemical activity of the ions for which they are specific and with which they are in equilibrium during the measurement. A typical example is the Ag/AgCl electrode. Other ion-specific electrodes usually consist of an Ag/AgCl electrode in a compartment of constant composition surrounded by a highly ion-permselective membrane (Fig. 15a). These electrodes are also reversible to the extent that the membrane is not leaky to ions of opposite sign. As they are unable to distinguish between the chemical and the electrical potential of the corresponding ion species two electrodes can be used to measure an electrical PD only if the chemical PD across the membrane is zero or known, and vice versa.

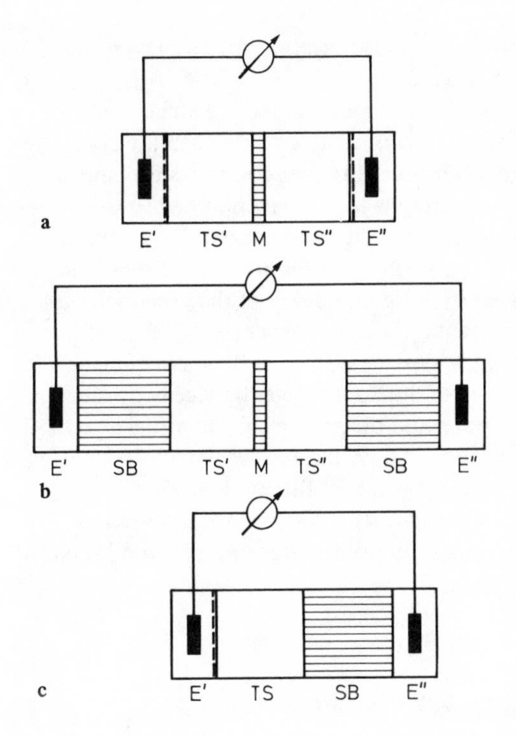

Fig. 15a-c. Equilibrium (reversible) and non-equilibrium (irreversible) electrodes. The difference between these two kinds of electrodes is schematically illustrated. **a** Two reversible, ion-specific electrodes (*E'*, *E"*) measuring the difference in electrochemical potential of the appropriate ion species between two test solutions (*TS'* and *TS"*) separated by a membrane (*M*). Ion-specificity is effectuated by "ion-sensing", i.e., strictly ion-selective membranes (*fat broken lines*). Note that the ion-sensing membranes are in immediate contact with the respective test solutions. **b** Two irreversible electrodes, applied to the same system as in **a**. The ion-selective membranes are replaced by salt bridges (*SB, shaded areas*) which separate the electrode chamber from the test solutions. In the ideal case the PD's at the phase boundaries (liquid junctions) between salt bridge and adjacent solutions are negligible so that the electrodes measure only the electrical PD between the test solutions across the membrane (*M*). **c** Combination of the two kinds of electrodes, both applied to the same test solution (*TS*), to measure the chemical PD of the ion species for which the reversible electrode is specific

To measure the electrical PD independently of ion concentrations, *irreversible electrodes* are usually preferred. These also employ a reversible electrode but the electrode compartment is separated from the test solution by a so-called salt bridge (Fig. 15b). The salt bridge usually contains the solution of an "inert" salt, e.g., KCl, which may or may not be identical with that of the electrode compartment. It is placed in a long capillary, in a porous clay, or in the meshwork of a gel (Agar-Agar) in order to prevent rapid intermixing with the adjacent solution at the liquid junctions. The salt bridge conducts electric current and hence favors electrical equilibration between electrode and test solution without appreciable movement of salt.

Owing to the liquid junctions there cannot be complete equilibrium between the electrode and the test solution as ions will slowly but continuously diffuse from the salt bridge into the adjacent solutions, leading to diffusion potentials at the boundaries ("liquid junction potentials"), for instance, between the bridge and the test solution and sometimes also between the bridge and the solution of the electrode compartment. Such boundary potentials are undefined but will add to the overall potential measurement, thereby inevitably introducing errors of unknown magnitude. One usually keeps these liquid junction potentials at a minimum by selecting for the bridge a salt consisting of ions which have almost the same mobilities, for instance KCl, so that the diffusion PD will be very small even if the bridge salt is present at high concentrations.

To measure the chemical activities of ions directly, a reversible and an irreversible (reference) electrode are brought into contact with the test solution. Both should measure the same electrical PD, which thus cancels, whereas only the specific, reversible electrode "senses" the ion concentration (Fig. 15c).

Microelectrodes have been applied to cells down to about 10 μm diameter. The results obtained with the giant squid axon, nitella, and other microorganisms are classic by now and need not be discussed here in detail. The use of similar electrodes for animal cells, on the other hand, even though many such results have been published since, is not of equally undisputed validity. Apart from the mentioned interference of liquid junction potentials, the membranes of animal cells appear to become easily damaged by the impalement of the electrode. In many instances a continuous decay of the potential was observed after the tip of the electrode had entered the cell, which may be taken as due to membrane damage. By greatly improving the time resolution of the measurement, Lassen et al. (1971) were able to reduce the error due to the potential decay by measuring the potential very shortly after the puncture, and the presumable potential at zero time is obtained by extrapolation, under the assumption that the decay had the same rate at the beginning as it had later. The values measured by this method were indeed considerably higher than those of previous measurements (Lassen and Rasmussen, 1978). Still it is likely that such measurements, in spite of their improvement, give too low values.

3.3 Distribution of Passive Permeant Ions

It has been discussed in a preceding section that passively permeant ions will in equilibrium have the same electrochemical potential on both sides of the membrane. Accordingly the following relation, usually referred to as the Nernst equation, must hold between the electrical PD and the distribution ratio of the ion:

$$\Delta\Psi = -\frac{RT}{zF} \ln \frac{c_i''}{c_i'}$$

z is the electrical valency of the ion i. Once such an ion has reached equilibrium distribution, or comes close to it, its distribution may be used as an indicator of the membrane potential, provided that the two concentrations in the adjacent solutions, c_i' and c_i'', are known. It has previously been explained that at static head, e.g., in a system with ion pumps, the passive ion will have reached such equilibrium distribution provided that the permeation of this ion is not coupled to any other permeation process. In other words, the flow of this ion must not be linked, by co- or counter-transport, by friction, or otherwise, to any other thermodynamically downhill flow of matter or energy through the membrane. The equilibrium distribution does not depend on the mobility of this ion, only the time it takes to reach this distribution.

In a *transient state*, by contrast, for instance after the pumps have been blocked by a metabolic inhibitor, the distribution of the passive and uncoupled ion is no longer in equilibrium but may come close to it, provided its penetration is rapid relative to that of the other ions in the system and not coupled to another permeation process. In other words, the degree to which the Nernst equation can be safely applied depends now very much on the mobility of the ion concerned: only passive, uncoupled, and very rapidly penetrating ions are useful for a PD measurement. The distribution of the passive Cl ions has long been considered a reliable indicator of the cellular transmembrane PD of muscle cells and others. In some cases, e.g., in Ehrlich cells, the PD measured by microelectrodes agrees well with that by chloride distribution (Lassen et al., 1971). Nevertheless, the Cl^- distributions is likely to be unreliable in this respect since none of the above-mentioned conditions may be met. Firstly, it is not certain that Cl-ions are homogeneously distributed in the whole cellular space. For Ehrlich cells, for instance, it has been derived from experiments with cellular fractionation that the nucleus sequesters Cl-ions, so that the cytoplasmic Cl^- concentration is likely to be much lower than the overall Cl^- concentration (Pietrzyk and Heinz, 1974). As a consequence, the cellular membrane potential should be underestimated by this method, as was probably true for the values obtained by microelectrodes. Secondly, the permeability of the cell membrane for Cl^- net movement is very much slower than previously assumed, the major part of Cl^- fluxes being accounted for by exchange diffusion. Hence the transport systems may not have reached complete steady state at the time of the measurement, and the Cl^- distribution may still be far from equilibrium. Thirdly, the Cl^- flux is likely to be coupled to the flow of other ions, e.g., HCO_3^- of OH^- via counter-transport, and of Na^+ and K^+ via cotransport. As a consequence Cl^- distribution may be held at disequilibrium by a stationary flow of these ions out of the cell (Heinz et al., 1977; Geck et al., 1980).

In view of these uncertainties concerning Cl^- distribution, other permeant ions have been suggested for PD measurement. Several lipid soluble ions, cations, and anions, are now available and have been tested in various cells and mitochondria (Bakeeva et al., 1970). Most of the cations are quaternary ammonium or phosphonium bases with alkyl and aryl substituents, such as dibenzyl-dimethyl ammonium (DDA), tetrabutyl ammonium (TBA), triphenylmethyl phosphonium (TPMP), or tetraphenyl phosphonium (TPP). The readiness to penetrate the membrane varies from compound to compound and from system to system, so that in each system it should be tested which of the indicator ions penetrate fast enough to serve the purpose, especially in transient systems. The penetration of these cations can in some cases be accelerated by added traces of the lipid soluble anion tetraphenyl boron (TPB). For cells and mitochondria, which are negative inside, cations are more suitable for PD measurement because of the higher distribution ratio. If the compartment to be tested is positive inside, such as inside-out vesicles or some artificial liposomes, lipid soluble anions are preferable for the same reason, such as the already mentioned tetraphenyl boron (TPB), or phenyldicarbaundecaborane (PDB). For a review see Skulachev (1971).

Many of the ions mentioned have been shown to be metabolic inhibitors; this undesired side effect can be avoided by using the isotopically labeled ions in tracer amounts. Whereas it is unlikely that these ions are, like Cl-ions sequestered in the cell nuclei, they may be bound to cellular material and are likely to be accumulated in active mitochondria inside the cell, so that erroneously high estimates of the PD of the cell membrane may result.

3.4 Fluorescent Dyes

A specific relationship has been assumed to exist between membrane PD and the fluorescence intensity of certain dyes. Such dyes, if applied to membrane systems, rapidly respond by a change in fluorescence intensity (or in the associated transmission) to a change in electrical potential across the membrane of a cell or vesicular structure (Hoffman and Laris, 1974). The method so far is largely empirical, since neither the quantitative relationship between PD and fluorescence, nor the molecular mechanism underlying the observed changes has been completely elucidated. In addition, the molecular mechanism does not appear to be the same for all fluorescent compounds so far tested. For these reasons, it is indispensable that each particular compound, for each system to which it is applied, be calibrated first on the basis of known electrical potential differences.

It seems that in this respect most potential-sensitive fluorescent dyes can be divided into two fundamentally different groups: the permeant ones and the impermeant ones. To the first group belong, among others, the cationic *cyanines* and the anionic *oxonols*. Though all of these dyes are ionic, they easily permeate a hydrophobic barrier, supposedly because their ionic groups are buried within a bulk of hydrophobic groups, so that the charge of the whole molecule appears to be "smeared" or delocalized. Hence, the primary response of these dyes to an electric potential is to distribute themselves between the two phases across the membrane according to the electrical potential, and the equilibrium distribution should be related to the electrical potential according to the Nernst equation. In this respect they should behave similarly to the previously discussed lipophilic ions. If, for instance, the potential across the compartment membrane increases, the inside being negative, (hyperpolarization) more of the cationic cyanine should be accumulated inside the cell. Somewhat more difficult is to answer the question why such an increase in accumulation leads to a decrease in fluorescence of cationic dyes. It seems that their fluorescence is quenched by the absorption to the cellular membrane (or to other cellular components) in which process the cyanine molecules dimerize. If for instance after "hyperpolarization" action the cationic dye accumulates inside the cell its absorption to or more strongly into the membrane will be enhanced. In this way substantial amounts of fluorescent dye are removed from free solution and quenched. If, on the other hand, the cell is depolarized, much of the same dye would come out of the cell, and a great amount of the fluorescent monomers would be redissolved. Even though the outside concentration of dye might rise somewhat under these conditions, the large excess of extracellular fluid will keep this rise in concentration low, so that the increase in adsorption to the membrane from the outside will be small as compared to that from the inside under the condition of hyperpolarization. With anionic dyes, such as oxonol, the opposite behavior of fluorescence would have to be expected during the same changes in electrical PD (Cohen and Salzberg, 1978; Waggoner, 1979).

Since in either case the final fluorescence after a change in PD is measured under equilibrium distribution of the dye, the quenching obtained with a given electrical PD should be the same, no matter whether the dye is initially present at the outside or at the inside of the cell.

In contrast to the cyanine and oxonol derivatives, the *merocyanine* derivatives do not appear to penetrate the membrane readily, because their electric charge is strictly localized and thus exposed to the environment. While the quenching associated with the appropriate changes in electrical PD may also be due to dimerization following absorption to the membrane, the relationship between quenching and potential change appears to be fundamentally different from those of the permeant dyes. This is because dye added from the outside can be adsorbed only to the outer part of the bilayer membrane, and that added from the inside only to the inner part. Accordingly, at the same electrical PD, the degree of quenching should be different if the dye is present at the outside from what it would be if the dye were present at the inside only, since the electrical charge of the membrane should be different inside and outside in the presence of an electrical transmembrane PD. For instance, in the case of hyperpolarization, a merocyanine, negative through its ionized sulfonic group, would easily adsorb to the (positive) outside of the cellular membrane but not to the inside, whereas in the case of depolarization the opposite should be expected.

With other PD indicator ions the permeant fluorescent dyes share the source of error due to compartmentalization, especially if the indicator is also accumulated in the mitochondria whose membranes have a PD of their own. Hence all these indicator ions have given the best results with one-compartment systems such as mitochondria-less bacteria or vesicles and with mitochondria themselves. The possible interference of compartmentalization might be a serious handicap, and little can be done about it at the present time.

The merocyanines have the advantage of responding much faster than the penetrating dyes (cyanines and oxonols), but one does not quite know yet whether and to what extent the response of the merocyanines is due to phase boundary potentials, and to what extent it can be used to determine the transmembrane potential difference. A further drawback is that the size of the signals as compared to that of the noise (signal: noise ratio) is so small that it is technically difficult to obtain significant results (Cohen and Salzberg, 1978; Waggoner, 1979).

3.5 Other Membrane Probes

Several fluorescents and other dyes, if embedded naturally or artificially in a membrane, have been shown to react to changes in the potential across this membrane. The mostly used dye of this kind is aminonaphthalene sulfonate (ANS) which is supposed to react to PD changes as well as to other, in particular structural changes within the membranes. This compound is widely used, mainly for the latter purpose. Other fluorescent dyes, such as acridine derivatives like atebrin, supposedly react also to pH differences across a membrane and are therefore not ideally suitable as PD indicators. Finally, there are some other dyes of the carotine and chlorophyll families which, if present in a membrane, respond by a change of their light absorption to a change of the PD. Only few of these have been practically used so far. Good results have apparently been obtained by such a substance in chloroplasts (electrochroism) with an absorption change at 515 nm, caused by a shift of absorbant bands in a transmembrane electrical field. This absorbant change has been found to be linearly related to the change in electrical potential difference (Gräber and Witt, 1976).

Acknowledgment. This manuscript was completed while I was holding a USPHS, NIH Grant (No. RO 1 GM 26554-01). I would like to thank Dr. B. Hess for his helpful criticism. I would also like to express my gratitude to Mrs. B. Pfeiffer for the skillful drawing of most of the figures, and to Mrs. Ruth Temple for aptly and speedily typing the manuscript.

References

Adrian, R.H., Slayman, C.L.: Membrane potential and conductance during transport of sodium, potassium and rubidium in frog muscle. J. Physiol. *184*, 970-1017 (1966)

Andersen, O.S., Finkelstein, A., Katz, I., Cass, A.: Effect of phloretin on the permeability of thin lipid membranes. J. Gen. Physiol. *67*, 749-771 (1976)

Armstrong, C.M., Binstock, L.: Anomalous rectification in the squid giant axon injected with tetraethylammonium chloride. J. Gen. Physiol. *48*, 859 (1965)

Armstrong, W.McD., Bixenman, W.R., Frey, K.F., Garcia-Diaz, J.F., O'Regan, M.G., Owens, J.L.: Energetics of coupled Na^+ and Cl^- entry into epithelial cells of bullfrog small intestine. Biochim. Biophys. Acta *551*, 207-219 (1979)

Bakeeva, L.E., Grinius, L.L., Jasaitis, A.A., Kuliene, V.V., Levitsky, D.D., Liberman, E.A., Severina, I.I., Skulachev, V.P.: Conversion of biomembrane-produced energy into electric form. 11. Intact mitochondria. Biochim. Biophys. Acta *216*, 13 (1970)

Bayer, R., Hennekes, R., Kaufmann, R., Mannholt, R.: Inotropic and electrophysiological actions of verapamil and D600 in mammalian myocardium. Naunyn Schmiedebergs Arch. Pharmacol. *290*, 49 (1975)

Bentley, P.J.: Amiloride a potent inhibitor of sodium transport across the toad bladder. J. Physiol. *195*, 317 (1968)

Berner, W., Kinne, R.: Transport of p-amino hippuric acid by plasma membrane vesicles isolated from rat kidney cortex. Pflügers Arch. *361*, 269-277 (1976)

Bockris, J.O'M., Reddy, A.K.N.: Modern Electrochemistry. New York: Plenum, 1973

Cabantchik, Z.I., Rothstein, A.: The nature of the membrane sites controlling anion permeability of human red blood cells as determined by studies with disulfonic stilbene derivatives. J. Membr. Biol. *10*,311 (1972)

Cabantchik, F.I., Rothstein, A.: Membrane proteins related to anion permeability of human red blood cells. J. Membr. Biol. *15*, 227-248 (1974)

Cass, A., Finkelstein, A., Krespi, V.: The ion permeability induced in thin lipid membranes by the polyene antibiotics nystatin and amphotericin B, J. Gen. Physiol. *56*, 100-24 (1970)

Chappell, J.B., Crofts, A.: Gramicidin and ion transport in isolated liver mitochondria. Biochem. J. *95*, 393-402 (1965)

Cohen, L.B., Salzberg, B.M.: Optical measurement of membrane potential. Rev. Physiol. Biochem. Pharmacol. *83*, 37-88 (1978)

Crane, R.K.: The gradient hypothesis and other models of carrier mediated active transport. Rev. Physiol. Biochem. Pharmacol. *78*, 101-159 (1977)

Diamond, J.M.: The mechanism of solute transport by the gallbladder. J. Physiol (London) *161*, 474-502 (1962)

Dilger, J.P., McLaughlin, St.F.A., McIntosh, T.J., Simon, S.A.: The dielectric constant of phospholipid bilayers and the permeability of membrane to ions. Science *206*, 1196 (1979)

Donnan, F.G.: Theorie der Membrangleichgewichte und Membranpotentiale bei Vorhandensein von nicht dialysierende Elektrolyte. Ein Beitrag zur physikalisch-chemischen Physiologie. Z. Elektrochem. *17*, 572-581 (1911)

Essig, A., Caplan, S.R.: Energetics of active transport processes. Biophys. J. *8*, 1434-1457 (1968)

Finkelstein, A., Andersen, O.S.: The gramacidin A channel. J. Membr. Biol. (1980) (in press)

Finkelstein, A., Mauro, A.: Equivalent circuits as related to ionic systems. Biophysical J. *3*, 215 (1963)

Finkelstein, A., Mauro, A.: Physical Principles and Formalisms of Electrical Excitability. Handbook of Physiology. The Nervous System, Vol. 1, American Physiological Society, Maryland. 161 (1977)

Geck, P., Pietrzyk, C., Burckhardt, B.C., Pfeiffer, B., Heinz, E.: Electrically silent cotransport of Na^+, K^+ and Cl^- in Ehrlich cells. Biochim. Biophys. Acta *600*, 432-447 (1980)

Goldman, D.E.: Potential, impedance and rectification in membranes. J. Gen. Physiol. *27*, 37 (1944)

Gräber, P., Witt, H.T.: Relations between the electrical potential, pH gradient, proton flux and phosphorylation in the photosynthetic membrane. Biochem. Biophys. Acta *423*, 141 (1976)

Graven, S.N., Estrada-O, S., Lardy, H.A.: Alkali metal cation release and respiratory inhibition induced by nigericin in rat liver mitochondria. Proc. Natl. Acad. Sci. USA *56*, 654 (1966)

Graven, S.N., Lardy, H.A., Estrada-O, S.: Antibodies as tools for metabolic studies: VIII. Effect of nonactin Homologs on alkali metal cation transport. Biochemistry *6*, 365-371 (1967)

Gunn, R.B., Cooper, J.A., Jr.: Effect of local anesthetics on chloride transport in erythrocytes. J. Membr. Biol. *25*, 311-326 (1975)

Hall, J.E.: Channels in black lipid films. In: Membrane Transport in Biology (eds. G. Giebisch, D. Tosteson, H.H. Ussing), Vol. 1, pp. 475-53. Berlin-Heidelberg-New York: Springer, 1978

Hall, J.E., Mead, C.A., Szabo, G.: A barrier model for current flow in lipid bilayer membranes. J. Membr. Biol. *11*, 75-97 (1973)

Harold, F.M.: Ion currents and physiological functions in microorganisms. Annu. Rev. Microbiol. *31*, 181-203 (1977)

Harold, F.M., Altendorf, K.H., Hirata, H.: Probing membrane transport mechanisms with ionophores. Ann. N.Y. Acad. Sci. *235*, 149 (1974)

Harold, F.M., Pavlasova, E., Baarda, J.R.: A trans membrane pH gradient in S. Faecalis. Origin and Dissipation by proton conductors and N, N'-dicyclohexylcarbodiimide. Biochem. Biophys. Acta *196*, 235-244 (1970)

Heinz, E.: Mechanics and Energetics of Biological Transport. Molecular Biology, Biochemistry and Biophysics, Vol. 29. Berlin-Heidelberg-New York: Springer, 1978

Heinz, E.: The Response of the "Protonmotive Force" to the pulse of an electrogenic proton pump. Curr. Top. Membr. Transp. *14* (1981) (in press)

Heinz, E., Geck, P.: The electrical potential difference as a driving force in Na^+-linked cotransport of organic solutes. In: Membrane Transport Processes, Vol. 1 (ed. J.F. Hoffman), pp. 13-30. New York: Raven Press, 1978

Heinz, E., Geck, P., Pietrzyk, C.: Driving forces of amino acid transport in animal cells. Ann. N.Y. Acad. Sci. *264*, 428-441 (1975)

Heinz, E., Geck, P., Pietrzyk, C., Pfeiffer, B.: Electrogenic ion pump as an energy source for active amino acid transport in Ehrlich cells. In: Biochemistry of Membrane Transport (eds. G. Semenza, E. Carafoli), pp. 236-249. Berlin-Heidelberg-New York: Springer, 1977

Henderson, P.: Zur Thermodynamik der Flüssigkeiten. Z. Phys. Chem. *59*, 118-127 (1907)

Henderson, F.J.F., McGivan, J.D., Chappell, J.B.: The action of certain antibodies on mitochondrial, erythrocyte and artificial phospholipid membranes. Biochem. J. *111*, 521-535 (1969)

Heyningen, R. van: Metabolism of xylose by the lens. 2. Rat lens in vivo and in vitro. Biochem. J. *73*, 197-207 (1959)

Heytler, P.G., Pritchard, W.W.: A new class of uncoupling agents: carbonyl cyanide phenyl hydrazones. Biochim. Biophys. Res. Commun. *7*, 272-275 (1962)

Hille, B.: Common mode of action of three agents that decrease the transient change in Na^+ permeability in nerves. Nature *210*, 1220 (1966)

Hodgkin, A.L.: The Croonian lecture: Ionic movements and electrical activity in giant nerve fibres. Proc. R. Soc. London Ser. B *148*, 1-37 (1957)

Hoffman, J.F., Laris, P.C.: Determination of membrane potentials in human and amphiuma red blood cells by means of a fluorescent probe. J. Physiol. *239*, 519 (1974)

Junge, W.: Physical aspects of electron transport and photophosphorylation in green plants. Ber. Dtsch. Bot. Ges. *88*, 283-301 (1975)

Kashket, E.R., Wilson, T.H.: Role of metabolic energy in the transport of B-galactosides by streptococcus lactis. J. Bacteriol. *109*, 784-789 (1972)

Katz, B.: Nerve, Muscle and Synapse. New York: McGraw-Hill 1966

Kedem, O., Caplan, S.R.: Degree of coupling and its relation to efficiency of energy conversion. Trans. Faraday Soc. *61*, 1897 (1965)

Kedem, O., Essig, A.: Isotope flows and flux rations in biological membranes. J. Gen. Physiol. *48*, 1047-1070 (1965)

Keynes, R.: The molecular organization of the sodium channels in nerve. In: Biochemistry of Membrane Transport (eds. G. Semenza, E. Carafoli), pp. 442-448. Berlin-Heidelberg-New York: Springer, 1977

Kilberg, M.S., Christensen, H.N.: Electron-transferring enzymes in the plasma membrane of the Ehrlich ascites tumor cell. Biochemistry *18*, 1525-1530 (1979)

Kinsella, J.L., Aronson, P.S.: Properties of the Na^+-H^+-exchange in renal microvillus membrane vesicles, (Am. J. Physiol.) 238 (1980) F461-7

Kirsch, G.E., Narahashi, T.: 3, 4-Diamino pyridine. A potent new potassium channel blocker. Biophys. J. *22*, 507-512 (1978)

Klingenberg, M.: Metabolic transport in mitochondria: Example for intracellular membrane function. Essay Biochem. *6*, 119-159 (1970)

Komor, B., Komor, E., Tanner, W.: Transport of a structly coupled active transport system into a facilitated diffusion system by nystatin. J. Membr. Biol. *17*, 231-238 (1974)

Lassen, U.V., Rasmussen, B.E.: Use of microelectrodes for measurement of membrane potentials. In: Membrane Transport in Biology (eds. G. Giebisch, A.C. Tosteson, H.H. Ussing), Vol. 1, pp. 169-203. Berlin-Heidelberg-New York: Springer, 1978

Lassen, U.V., Nielsen, A.M., Pape, L., Simosen, L.O.: The membrane potential of Ehrlich ascites tumor cells – microelectrode measurements and their critical evaluation. J. Membr. Biol. *6*, 269 (1971)

Läuger, D., Neumcke, B.: Theoretical analysis of ion conductance in lipid bilayer membranes. In: Membranes, Vol. 2 (ed. G. Eisenman). pp. 1-59. New York: Dekker, 1973

LeBlanc, O.H., Jr.: The effect of uncouplers of oxidative phosphorylation on lipid bilayer membranes. J. Membr. Biol. *4*, 227-251 (1971)

Liedke, C.M., Hopfer, U.: Anion transport in brush border membranes isolated from rat small intestine. Biochem. Biophys. Res. Commun. *76*, 579-585 (1977)

Machen, T.E., Forte, J.G.: Gastric Section. In: Membrane Transport in Biology (eds. G. Giebisch, D.C. Tosteson, H.H. Ussing), Vol. IV B, pp. 643-748. Berlin-Heidelberg-New York: Springer, 1979

McLaughlin, S.: Electrostatic potentials at membrane solution interfaces. Curr. Top. Membr. Transp. *9*, 71-144 (1977)

Meyer, K.H., Sievers, J.F.: Permeability of membranes. I. Theory of ionic permeability. Helv. Chim. Acta *19*, 649-664 (1936)

Mitchell, P.: Coupling of phosphorylation to electron and hydrogen transfer by a chemiosmotic type of mechanism. Nature *191*, 144-8 (1961)

Mitchell, P.: Chemiosmotic Coupling and Energy Transduction. Bodmin: Glynn Research, 1966

Mitchell, P.: Chemiosmotic Coupling and Energy Transduction. Bodmin: Glynn Research, 1968

Mitchell, P., Moyle, J.: Acid-base titration across the membrane system of rat-liver mitochondria. Catalysis by uncouplers. Biochem. J. *104*, 588 (1967)

Moore, C., Pressman, B.C.: Mechanism of action of valinomycin on mitochondria. Biochem. Biophys. Res. Commun. *15*, 562-567 (1964)

Mueller, R.U., Finkelstein, A.: The effect of surface charge on the voltage dependent conductance induced in thin lipid membranes by monazomycin. J. Gen. Physiol. *60*, 285-306 (1972)

Mueller, P., Rudin, D.O.: Translocations in bimolecular lipid membranes: Their role in dissipative and conservative bioenergy transductions. Curr. Top. Bioenerg. *3*, 157 (1969)

Mullins, L.J.: Ion selectivity of carriers and channels. Biophys. J. *15*, 921 (1975)

Mullins, L.J.: A mechanism for Na^+/Ca^{++} transport. J. Gen. Physiol. *70*, 681-695 (1977)

Mullins, L.J., Awad, M.Z.: The control of the membrane potential of muscle fibers by the sodium pump. J. Gen. Physiol. *48*, 761 (1965)

Mullins, L.J., Noda, K.: The influence of sodium-free solutions on the membrane potential of frog muscle fibers. J. Gen. Physiol. *47*, 117 (1963)

Murer, H., Kinne, R.: The use of isolated membrane vesicles to study epithelial transport processes. J. Membr. Biol. *55*, 81-95 (1980)

Murer, H., Hopfer, U., Kinne, R.: Sodium/proton antiport in brush border membrane vesicles isolated from rat small intestine and kidney. Biochem. J. *154*, 597-604 (1976)

Nernst, W.: Zur Kinetik der in Lösung befindlichen Körper. Z. Phys. Chem. *2*, 613-637 (1888)

Okada, P., Halvorson, H.O.: Uptake of a-thioethyl D. glucopyranoside by saccharomyces cerevisiae. II. General characteristics of an active transport system. Biochim. Biophys. Acta *82*, 547 (1964)

Ovchinnikov, Y.A., Ivanov, V.T.: Recent developments in the structure-functional studies of peptide ionophores. In: Biochemistry of Membrane Transport (eds. G. Semenza, E. Carafoli), pp. 123-146. Berlin-Heidelberg-New York: Springer, 1977

Overbeck, J.T.G.: The donnan equilibrium. Progr. Biophys. *6*, 57-83 (1956)

Parlin, R.B., Eyring, H.: Membrane permeability and electrical potential. In: Ion Transport across Membranes (ed. H.T. Clarke), pp. 103-118. New York: Academic Press, 1954

Pietrzyk, C., Heinz, E.: The sequestration of Na^+, K^+ and Cl^- in the cellular nucleus and its energetic consequences for the gradient hypothesis of amino acid transport in Ehrlich cells. Biochim. Biophys. Acta *352*, 397-411 (1974)

Pietrzyk, C., Geck, P., Heinz, E.: Regulation of the electrogenic ($Na^+ + K^+$)-pump of Ehrlich cells by intracellular cation levels. Biochim. Biophys. Acta *513*, 89-98 (1978)

Planck, M.: Über die Potentialdifferenz zwischen zwei verdünnten Lösungen binaerer Elektrolyte. Ann. Phys. Chem., N.F., *40*, 561 (1890)

Pressman, B.C.: Ionophorous antibiotics as models of biological transport. Federation Proc. *27*, 1283-1288 (1968)

Pressman, B.C., deGuzman, N.T.: Biological applications of ionophores: theory and practice. Ann. N.Y. Acad. Sci. *264*, 373-386 (1975)

Racker, E.: Mechanism of ion transport and ATP formation. In: Membrane Transport in Biology (eds. G. Giebisch, D.L. Tostoson, H.H. Ussing), pp. 259-90. Berlin-Heidelberg-New York: Springer, 1978

Reed, P.W., Lardy, H.A.: A 23187; A divalent cation ionophore. J. Biol. Chem. *247*, 6970-6977 (1972)

Rehm, W.S.: Models for electrogenic proton transport mechanisms. Ann. N.Y. Acad. Sci. *341*, 1-11 (1980)

Ritchie, J.M.: Electrogenic ion pumping in nervous tissue. Curr. Top. Bioenerg. *4*, 327-356 (1971)

Ritchie, J.M., Rogart, R.B.: The binding of saxitoxin and tetrodotoxin in excitable tissue. Rev. Physiol. Biochem. Pharmacol. *79*, 1-50 (1977)

Rothstein, A., Cabantchik, Z.I., Knauf, P.: Mechanism of anion transport in red blood cells: role of membrane proteins. Fed. Proc. *35*, 3-10 (1976)

Rottenberg, H.: The thermodynamic description of enzyme-catalized reactions. Biochem. J. *13*, 503 (1973)

Rottenberg, H.: The measurement of transmembrane electrochemical proton gradients. Bioenergetics *7*, 61-74 (1975)

Rottenberg, H.: An irreversible thermodynamic approach to energy coupling in mitochondria and chloroplasts. In: Progress in Surface and Membrane Science (eds. J.F. Danielli, A. Cadenhead). Vol. 12, p. 245. New York: Academic Press, 1978

Sachs, G.: H^+-Transport by a non-electrogenic gastric ATpase as a model for acid secretion. Rev. Physiol. Biochem. Pharmacol. *79*, 133-180 (1977)

Schlögl, R.: Elektrodiffusion in freier Lösung und geladenen Membranen. Z. Physk. Chem. *1*, 305 (1954)

Schlögl, R.: Stofftransport durch Membranen. Darmstadt: Steinkopff, 1964

Selwyn, M.J., Dawson, A.P., Stockdale, M., Gains, N.: Chloride-hydroxide exchange across mitochondrial erythrocyte and artificial lipid membranes mediated by trialkyl and triphenyl tin compounds. Eur. J. Biochem. *14*, 120-126 (1970)

Sen, A.K., Post, R.L.: Stoichiometry and localization of adenosine triphosphate-dependent sodium and potassium transport in the erythrocyte. J. Biol. Chem. *239*, 345 (1964)

Simon, W., Proda, L.A.R., Wipf, H.K.: Cation specificity of inhibitors. In: Inhibitors. Tools in Cell Research (eds. Th. Bücher, H. Sies), 20. Coll. Ges. Biol. Chem., pp. 356-364. Berlin-Heidelberg-New York: Springer, 1969

Skulachev, V.P.: Energy transformations in the respiratory chain. Curr. Top. Bioenerg. *4*, 127 (1971)

Slayman, C.L., Long, W.S., Lu, C.Y.-H.: The Relationship between ATP and an electrogenic pump in the plasma membrane of neurospora crassa. J. Membr. Biol. *14*, 3905 (1973)

Spencer, R.L., Lehninger, A.L.: L-lactate transport in Ehrlich ascites tumor cells. Biochem. J. *154*, 405 (1976)

Stark, G.: Monensin: a new biologically active compound produced by fermentation. Fermentation Adv. New York: Academic Press. Symp., 3 (1969) 517-540

Stark, G.: Carrier-mediated ion transport across thin lipid membranes. In: Membrane Transport in Biology (eds. G. Giebisch, D.C. Tosteson, H.H. Ussing), pp. 447-473. Berlin-Heidelberg-New York: Springer, 1978

Stoeckenius, W., Lozier, R.H., Bogomolni, R.A.: Bacteriorhodopsin and the purple membrane of halobacteria. Biochim. Biophys. Acta *505*, 215 (1979)

Teorell, T.: Zur quantitativen Behandlung der Membranpermeabilität. Z. Elektrochemie angew. physikal. Chemie *55*, 460 (1951)

Thomas, R.C.: Electrogenic sodium pump in nerve and muscle cells. Am. Physiol. Soc. *52*, 563 (1972)

Urry, D.W., Long, M.M., Jacobs, M., Harris, R.D.: Conformation and molecular mechanisms of carriers and channels. Ann. N.Y. Acad. Sci. *264*, 203-220 (1975)

Vidaver, G.A.: Mucate inhibition of glycine entry into pigeon red cells. Biochemistry *3*, 799 (1964a)

Vidaver, G.A.: Some tests of the hypothesis that the sodium ion gradient furnishes the energy for glycine-active transport by pigeon red cells. Biochemistry *3*, 803 (1964b)

Villereal, M.L., Cook, J.S.: Regulation of active amino acid transport by growth-related changes in membrane potential in a human fibroblast. J. Biol. Chem. *253*, 8257-8262 (1978)

Villereal, M.L., Levinson, C.: Chloride-stimulated sulfate efflux in Ehrlich ascites tumor cells: evidence for 1:1 coupling. J. Cell. Physiol. *90*, 553 (1977)

Waggoner, A.S.: Dye indicators of membrane potential. Annu. Rev. Biophys. Bioeng. *8*, 47 (1979)

Westerhoff, H.V., Dam, K. van: Irreversible thermodynamic description of energy transduction in biomembranes. Curr. Top. Bioenerg. *9*, 1-62 (1979)

Westerhoff, H.V., Scholte, B.J., Hellingwert, K.J.: Bacteriorhodopsin in liposomes. I. A description using irreversible thermodynamics. Biochim. Biophys. Acta *547*, 544-560 (1979)

Subject Index

Molecular Biology, Biochemistry and Biophysics

Editors:
A. Kleinzeller, G. F. Springer, H. G. Wittmann

Volume 1
J. H. van't Hoff
Imagination in Science
1967. 1 portrait. VI, 18 pages
ISBN 3-540-03933-3

Volume 3
T. Robinson
The Biochemistry of Alkaloids
1968. 37 figures. X, 149 pages
ISBN 3-540-04275-X

Volume 5
B. Jirgensons
Optical Activity of Proteins and Other Macromolecules
2nd revised and enlarged edition. 1973.
71 figures. IX, 199 pages. (The title of the first edition was: "Optical Rotatory Dispersion of Proteins and Other Macromolecules")
ISBN 3-540-06340-4

Volume 6
F. Egami, K. Nakamura
Microbial Ribonucleases
1969. 5 figures. IX, 90 pages
ISBN 3-540-04657-7

Volume 8
Protein Sequence Determination
A Sourcebook of Methods and Techniques
Editor: S. B. Needleman
2nd revised and enlarged edition. 1975.
80 figures. XVIII, 393 pages
ISBN 3-540-07256-X

Volume 9
R. Grubb
The Genetic Markers of Human Immunoglobulins
1970. 8 figures. XII, 152 pages
ISBN 3-540-05211-9

Volume 10
R. J. Lukens
Chemistry of Fungicidal Action
1971. 8 figures. XIII, 136 pages
ISBN 3-540-05405-7

Volume 11
P. Reeves
The Bacteriocins
1972. 9 figures. XI, 142 pages
ISBN 3-540-05735-8

Volume 12
T. Ando, M. Yamasaki, K. Suzuki
Protamines
Isolation, Characterization, Structure and Function
1973. 24 figures, 17 tables. IX, 114 pages
ISBN 3-540-06221-1

Volume 13
P. Jollès, A. Paraf
Chemical and Biological Basis of Adjuvants
1973. 24 figures, 41 tables. VIII, 153 pages
ISBN 3-540-06308-0

Volume 14
Micromethods in Molecular Biology
Editor: V. Neuhoff
With contributions by numerous experts
1973. 275 figures (2 in color), 23 tables.
XV, 428 pages
ISBN 3-540-06319-6

Volume 15
M. Weissbluth
Hemoglobin
Cooperativity and Electronic Properties
1974. 50 figures. VIII, 175 pages
ISBN 3-540-06582-2

Volume 16
S. Shulman
Tissue Specificity and Autoimmunity
1974. 32 figures. XI, 196 pages
ISBN 3-540-06563-6

Volume 17
Y. A. Vinnikov
Sensory Reception
Cytology, Molecular Mechanisms and Evolution
1974. 124 figures (173 separate illustrations).
IX, 392 pages
ISBN 3-540-06674-8

Springer-Verlag
Berlin
Heidelberg
New York

Volume 18
H. Kersten, W. Kersten

Inhibitors of Nucleic Acid Synthesis

Biophysical and Biochemical Aspects
1974. 73 figures. IX, 184 pages
ISBN 3-540-06825-2

Volume 19
M. B. Mathews

Connective Tissue

Macromolecular Structure and Evolution
1975. 31 figures. XII, 318 pages
ISBN 3-540-07068-0

Volume 20
M. A. Lauffer

Entropy-Driven Processes in Biology

Polymerization of Tobacco Mosaic Virus Protein and Similar Reactions
1975. 90 figures. X, 264 pages
ISBN 3-540-06933-X

Volume 21
R. C. Burns, R. W. F. Hardy

Nitrogen Fixation in Bacteria and Higher Plants

1975. 27 figures. X, 189 pages
ISBN 3-540-07192-X

Volume 22
H. J. Fromm

Initial Rate Enzyme Kinetics

1975. 88 figures, 19 tables. X, 321 pages
ISBN 3-540-07375-2

Volume 23
M. Luckner, L. Nover, H. Böhm

Secondary Metabolism and Cell Differentiation

1977. 52 figures, 7 tables. VI, 130 pages
ISBN 3-540-08081-3

Volume 24

Chemical Relaxation in Molecular Biology

Editors: J. Pecht, R. Rigler
With contributions by numerous experts
1977. 141 figures, 50 tables. XVI, 418 pages
ISBN 3-540-08173-9

Volume 25

Advanced Methods in Protein Sequence Determination

Editor: S. B. Neddleman
With contributions by numerous experts
1977. 97 figures, 25 tables. XII, 189 pages
ISBN 3-540-08368-5

Volume 26
A. S. Brill

Transition Metals in Biochemistry

1977. 49 figures, 18 tables. VIII, 186 pages
ISBN 3-540-08291-3

Volume 27

Effects of Ionizing Radiation on DNA

Physical, Chemical and Biological Aspects
Editors: A. J. Bertinchamps, J. Hüttermann, W. Köhnlein, R. Téoule
With contributions by numerous experts
1978. 74 figures, 48 tables. XXII, 383 pages
ISBN 3-540-08542-4

Volume 28
A. Levitzki

Quantitative Aspects of Allosteric Mechanisms

1978. 13 figures, 2 tables. VIII, 106 pages
ISBN 3-540-08696-X

Volume 29
E. Heinz

Mechanics and Energetics of Biological Transport

1978. 35 figures, 3 tables. XV, 159 pages
ISBN 3-540-08905-5

Volume 31

Membrane Spectroscopy

Editor: E. Grell
With contributions by numerous experts
1981. 146 figures. XI, 498 pages
ISBN 3-540-10332-5

Volume 32

Chemical Recognition in Biology

Editors: F. Chapeville, A.-L. Haenni
With contributions by numerous experts
1980. 210 figures, 39 tables. XX, 430 pages
ISBN 3-540-10205-1

Springer-Verlag
Berlin
Heidelberg
New York